T0214931

Immigration in the Circumpolar North

Immigration in the Circumpolar North: Integration and Resilience explores interconnected issues of integration and resilience among both immigrants and host communities in the Arctic region. It examines the factors that inhibit or enable the success of immigrants to the Arctic and the role of territoriality in the process of integration.

This book showcases a variety of perspectives on circumpolar immigration and includes insights from eight Arctic countries as well as thirteen 'observer countries' such as China, India, Singapore, Poland, Germany, France and Japan. It considers the solidarities and engagements of indigenous and other local peoples with the newly arrived immigrants and refugees, and the impact of immigration on the economic and societal life in the Circumpolar Arctic.

The book will be of interest to researchers, teachers, professors, policymakers and others interested in migration issues, Arctic issues, international relations, law and economic integration.

Nafisa Yeasmin is a doctor of Social Science and a post-doctoral researcher at the Arctic Centre of the University of Lapland, Finland. Dr Yeasmin works under the Arctic Governance Research Group. She has been leading the UArctic thematic network on Arctic Migration. She is a distinguished member of the Finnish National Ethnic Advisory Board and has been the president of the Arctic Immigrant Association.

Waliul Hasanat, LLD, is a professor of Law Discipline in Khulna University, Bangladesh. He studied for the Doctor of Laws at the Faculty of Law in the University of Lapland, on the Arctic Council. He was a post-doctoral fellow of the China–Nordic Arctic Research Centre in the School of Law and Political Science at the Ocean University of China.

Jan Brzozowski is an associate professor at Cracow University of Economics, Poland. His research interests include research methods in business and economics, business ethics, development of Latin America, international relations, economics of international migration, international migration, migration and socio-economic development.

Stefan Kirchner is an associate professor at the University of Lapland, Finland. He has specialized in international law. He has extensive practical experience in international law, having worked in both private practice and public administration.

Immigration in the Circumpolar North

Integration and Resilience

Edited by Nafisa Yeasmin,
Waliul Hasanat, Jan Brzozowski
and Stefan Kirchner

Routledge
Taylor & Francis Group

LONDON AND NEW YORK

First published 2021
by Routledge
2 Park Square, Milton Park, Abingdon, Oxon OX14 4RN

and by Routledge
605 Third Avenue, New York, NY 10017

First issued in paperback 2022

Routledge is an imprint of the Taylor & Francis Group, an informa business

British Library Cataloguing-in-Publication Data
A catalogue record for this book is available from the British Library

Library of Congress Cataloging-in-Publication Data
Names: Yeasmin, Nafisa, editor | Hasanat, Waliul, editor. | Brzozowski, Jan, editor. | Kirchner, Stefan, 1976– editor.
Title: Immigration in the circumpolar north : integration and resilience / edited by Nafisa Yeasmin, Waliul Hasanat, Jan Brzozowski and Stefan Kirchner.
Description: Abingdon, Oxon ; New York, NY : Routledge, 2020. | Includes bibliographical references and index.
Identifiers: LCCN 2020006747 (print) | LCCN 2020006748 (ebook)
Subjects: LCSH: Immigrants—Arctic regions—Social conditions. | Immigrants—Cultural assimilation—Arctic regions. | Indigenous peoples—Arctic regions—Social conditions. | Arctic regions—Emigration and immigration. | Arctic regions—Ethnic relations.
Classification: LCC JV9472 .I66 2020 (print) | LCC JV9472 (ebook) | DDC 304.80911/3—dc23
LC record available at https://lccn.loc.gov/2020006747
LC ebook record available at https://lccn.loc.gov/2020006748

ISBN: 978-0-367-36169-3 (hbk)
ISBN: 978-0-367-50531-8 (pbk)
ISBN: 978-0-429-34427-5 (ebk)

DOI: 10.4324/9780429344275

Typeset in Galliard
by Swales & Willis, Exeter, Devon, UK

Contents

Figures

Contributors

Ria-Maria Adams is a PhD candidate in Social and Cultural Anthropology at the Faculty of Social Sciences at the University of Vienna. Her research interests revolve around Arctic youth well-being, with a particular focus on immigrant youth and their cultural adaptation processes. She is funded by the Uni:docs fellowship programme of the University of Vienna and contributes, as a guest researcher at the University of Lapland, to the project: 'Live, Work or Leave? Youth – wellbeing and the viability of (post) extractive Arctic industrial cities in Finland and Russia'.

Jan Brzozowski is Head of the Department of European Studies and Economic Integration at Cracow University of Economics in Poland. He was a post-doctoral fellow in the Department of Economic and Social Sciences at the Università Politecnica delle Marche in Ancona. He is also a member of the Committee for Migration Research of the Polish Academy of Sciences. He specializes in economics of international migration, including such topics as economic integration, migrant transnationalism, and immigrant entrepreneurship.

Kheirie El Hariri graduated in Spring 2019 from the University of Akureyri with a Master of Arts by research from the faculty of Social Sciences. She has experience in teaching Syrian refugee students at elementary level in Lebanon and in mentoring teachers dealing with students coming from difficult backgrounds, such as refugees or students from low socio-economic classes.

Robert FitzSimmons is a lecturer at the University of Lapland, Finland. He has published numerous articles in peer-reviewed journals such as *Critical Education* and *Journal of Critical Education Policy Studies*. His major focus is on critical pedagogy and its use in classroom practice. He has been a teacher for more than four decades.

Hermína Gunnþórsdóttir is a professor at the University of Akureyri. She holds a BA degree in Icelandic and a teaching certificate from the University of Iceland, a Master's degree from the Iceland University of Education (2003) and a PhD from the University of Iceland (2014). She has worked in kindergarten, primary and secondary schools. Her teaching and research interests are related to inclusive school and education, multiculturalism and education, social justice in education, disability studies, and educational policy and practice.

Waliul Hasanat, LLD, is Dean of the Law School of Khulna University, Bangladesh. He is a visiting researcher in the Arctic Governance Research Group in the Arctic Centre of the University of Lapland, Finland. He is a Doctor of Laws from the Faculty of Law in the University of Lapland. He has published a number of articles and a few books with an emphasis on Arctic governance. He was a post-doctoral fellow of the China–Nordic Arctic Research Centre in the School of Law and Political Science of the Ocean University of China.

Stefan Kirchner is Associate Professor of Arctic Law at the Arctic Centre in the University of Lapland in Rovaniemi, Finland, from where he also gained a post-doctoral degree of *dosentti* (adjunct professor) in Fundamental and Human Rights. He has taught international law and human rights at universities in Finland, Germany, Italy, Lithuania, Ukraine and Greenland and has been a lawyer in private and government practice in Germany, specializing in public international law and human rights law, and in particular human rights litigation at the European Court of Human Rights and Germany's Federal Constitutional Court.

Timo Koivurova has a multidisciplinary specialization in Arctic law and government but has also conducted broader research on global law (including authoring an English-language textbook on international environmental law). Some of his research areas include: the legal status of the indigenous peoples of the Arctic (e.g. the Sámi), Arctic immigration and the continued development of the Arctic Council as an intergovernmental forum. He has been involved as an expert in several international processes globally and in the Arctic region and has published on the above-mentioned topics extensively.

Markus Meckl holds a PhD from Berlin Technical University where he studied at the Center for Research on Anti-Semitism. Since 2004, he has been working at the University of Akureyri in Iceland, where he is a professor in the School of Humanities and Social Sciences. In recent years, one of his research interests has focused on immigration issues in Iceland, as a result of which a series of articles has been published. Drawing on a broad range of experience including research at Riga University and the Latvian Academy of Culture, he is involved in immigration- and integration-related projects in the Nordic and Baltic countries.

Amelia Merhar is a Human Geography PhD candidate at the University of Waterloo. She has a Master's degree from York University, a Bachelor's degree from the University of Alaska Fairbanks and a Northern Studies Diploma from Yukon College. Her research interrogates the long-term health and relational consequences of systemic displacement, residential instability and hypermobility in Canada. She is the Director of the Crawford Research Institute, has been a research associate at the Robarts Centre for Canadian Studies, and worked as an analyst for the Council of Yukon First Nations.

Susanna Pääkkölä, MSc, is a visiting researcher at the Arctic Centre of the University of Lapland in Rovaniemi, Finland. She is a doctoral candidate in Biomedicine and is working on human thermoregulation in cold environments for her PhD dissertation research in the University of Oulu, Finland. Her biomedical research provides information about the thermoregulatory responses in persons with disabilities, the product safety of touristic outdoor activities and working conditions in cold climates.

Juha Suoranta is a tenured professor of Adult Education at the University of Tampere, Finland. In his career he has worked as a visiting scholar at the University of Illinois at Urbana-Champaign (1996–1997) and UCLA (2003–2004), and as a visiting professor at the University of Minnesota (2005–2006). He has published numerous articles and books on critical pedagogy including 'Radikaali kasvatus' ('Radical education', 2005), 'Taisteleva tutkimus' ('Rebellious research methods', 2014), 'C. Wright Millsin sosiologinen elämä' ('C. Wright Mills's sociological life', 2017) and 'Paulo Freire, sorrettujen pedagogi' ('Paulo Freire, a pedagogue of the oppressed', 2019).

Satu Uusiautti, PhD, is the Vice-Rector and Professor of Education at the University of Lapland, Finland. Her research interests focus on positive psychology, leadership and success, especially in work contexts. She has also researched positive development and flourishing at various phases of life and in educational contexts.

Nafisa Yeasmin is a doctor of Social Sciences and is a researcher on International Relations and Politics at the University of Lapland, Finland. Her main research interest focuses on immigration to the North – for example, sustainable entrepreneurship development, socio-economic integration of immigrants in the Arctic, regional development, community sustainability, social inclusion and migration governance. Recent interests include the well-being of immigrant youth and integrated education. She is leading the UArctic Thematic Network on Arctic Migration.

Part I

Introduction

1 Introduction

Migration and ethnic challenges for the Circumpolar North

Nafisa Yeasmin, Waliul Hasanat, Jan Brzozowski and Stefan Kirchner

The Circumpolar North is back in the spotlight. After the relative decline in the world's public interest after the end of the Cold War confrontation, the big powers are back in the game for the Arctic's natural resources and its strategic routes. People might perceive ideas such as Donald Trump's proposal to purchase Greenland as not serious, but this idea is simply a manifestation of a new, serious trend: the rising interest of the United States, Russia, Canada, Sweden, Finland, Norway and also non-Arctic states such as China to exert greater control over these territories. With global warming and a simultaneous technological and transportation revolution, the accessibility of this formerly remote and unfriendly region is gradually increasing, which attracts potential new settlers from various parts of the world. At the same time, demographic pressures on other continents are staggering: in Africa alone, the population is expected to rise to 1.68 billion by 2030, an increase of 42 percent compared with 2015 (United Nations, 2015). Even if we assume that conservative (or even restrictive) migratory regimes would prevail in the northern hemisphere in the coming decades, it is reasonable to expect that many migrants from the densely populated areas of Asia, Latin America or Africa would look for new opportunities, also settling in the Circumpolar North. Therefore, the main aim of this volume is to focus on migration to, and ethnic challenges in, the Circumpolar North, and to analyse the social, economic, political and cultural changes induced by population movements in this region.

Indeed, the northern parts of the world

> have been [already – editors' comment] receiving larger number[s] of im-migrants since the mid-80's, however with hundreds of thousands of people migrating and seeking asylum all over Europe, this was something new. As many were determined to settle in Northern Europe, the Arctic region had to face newfound problems and challenges [alongside] the already existing ones of Arctic migration. Cultural differences, [the] acceptance of foreign cit-izens within local communities, integration into local communities and [the] labour market are some of the issues [that] need co-operation in order to contribute to innovative solutions.
>
> (UArctic, 2017)

Diversity has a profound impact on the quality of life in the Arctic, which leads to a debate and the need to recognise and address the issues for determining the resilient capacity of the Arctic and polar people (UArctic, 2017).

In this framework, the concept of community resilience becomes a focal point in analyses of the socio-cultural and economic integration of newcomers in the Arctic and also in the process of negotiating and designing sound public policies that could support these processes. Community resilience is a concept that could be understood as a readiness to react in a positive and constructive way to social and economic transformation, but at the same time being able to preserve local cultural and social values and systems (Berkes and Ross, 2013). This implies that the "natives" should be willing to receive and appreciate the new ethnic groups that arrive in their "homeland" territory. They should positively value their norms, habits, modes of behaviour and religious beliefs, but at the same time pursue activities aimed at preserving local traditions and costumes. However, increasing the capacity for being resilient also needs certain kinds of determinants from immigrants. Immigration is a new phenomenon for the Circumpolar North. Community resilience can partially overcome the risk of exclusion and can enhance local control on integration and collective efficacy.

The integration of immigrants into a new socio-ecological system is challenging. The human wealth and resilience of immigrants affect, influence and control various actions and interactions (Yeasmin, 2018). The well-being of immigrants requires a positive environment that is an unbounded source of resources fulfilling human needs and supports the integration of immigrants and the emergence of self-efficacy (Yeasmin, 2018). The inclusion of immigrants in the mainstream Arctic ecological system demands evolving social dynamics that can encompass understanding and the acquisition of knowledge and incorporate the emergence of adaptive management as well as governance between actors and interest groups for a positive paradigm of integration (Yeasmin, 2018).

This book will highlight the models, metaphors and measures of Arctic resilience by exploring the multilevel factors that can hinder or enable the social inclusion of immigrants in the Arctic. This book will emphasise all of the discourses and paradoxes that control social inclusion in the Arctic. It is also important to explore how those modes of discourse regulate the integration regime. Identifying the integration regime of Arctic welfare states to secure settlement policies and build the community resilience of minorities indeed raises challenges and obligations both for locals and immigrants, and exploring opportunities and potential from both perspectives is equally important for the move towards an integrated Arctic society. The book will elaborate on discussions about contentious issues that affect socio-spatial learning in the changing Arctic. The Arctic is a sparsely populated area and thus it requires mutual understanding between communities for a nexus framework of integration and community spirit, which is one of the drivers for the development of such a territory. This discussion is undeniably applicable to and equally worthwhile for the integration of immigrants and sustainable community development. This book will bring different perspectives from different parts of the Arctic. The structure of our monograph is as follows.

Chapter 2, written by Dr Nafisa Yeasmin and Dr Satu Uusiautti, gives an excellent comparative analysis of two countries that are top-scoring in the Programme for International Students Assessment (PISA) tests. How diversity and superdiversity impact the educational system of these countries is discussed in this chapter. Singapore and Finland, two unique countries whose ethnic diasporas struggle with their national identities, have performed well in the PISA tests but have quite different educational systems and immigration structures. The authors explore how two diverse educational systems manage their educational integrity and the integration of children into school as well as into the society. A superdiverse society can generate a discursive argument on both the positive and negative impacts of integration and the social resilience of children in school that can either help or hinder understanding of the respective society from diverse perspectives. Learning styles are unique and distinctive, and they just need a logical approach and concept of learning that can facilitate the educational system regardless of whether or not it is a diverse or superdiverse country.

Chapter 3, by Kheirie El Hariri, Hermína Gunnþórsdóttir, and Markus Meckl, deals with the novel, but also politically sensitive, topic of Syrian refugees and their integration into the Arctic countries. The contributors present a case study of Syrian children and their parents from the perspective of the Icelandic educational system. Their qualitative study demonstrates how challenging it can be to find common ground in cases where the cultural differences between hosts and guests are vast; both sides, in this regard, are quite far from the desired example of community resilience. In the case of refugees, a problem that is quite common for all the Arctic region is a serious challenge to further integration: the feeling of temporariness and the clear intention to move from Iceland to English-speaking countries. But also, the "liberal" approaches in the educational system that are so emblematic of Scandinavian countries, when faced with the reality of "cultural otherness", present quite conservative and unfriendly features. The authors demonstrate that, although liberal, Icelandic public schools lack flexibility towards newcomers' needs and sensitivity towards their current problems. This case study offers interesting insights into other Arctic countries that have received and are still receiving considerable inflows of politically driven migrants.

Chapter 4 on youth immigrants, written by Ria-Maria Adams, deals with understanding the opportunities and challenges of youth immigrants as experienced in the Arctic, stemming mainly from their struggles to cope with a new environment with an extreme climate. It examines a few key factors that impact the well-being of immigrant youth from both theoretical and factual perspectives. The theoretical analysis involves concepts such as "diversity" and "superdiversity" and how they can be applicable in the Arctic, with special reference to Steven Vertovec, an anthropologist dealing with issues concerning ethnic and religious minorities. The core of Vertovec's arguments supports focusing on anthropology matters while conducting research on youth immigrants residing in the Arctic. The author describes the experiences of immigrant youth in the Arctic, in particular in Finnish Lapland, where they enjoy a high standard of living and quality services that are ensured by the government of a developed welfare state; in contrast, there is frustration at prevalent discrimination in obtaining employment and integrating into the mainstream society. The significance of immigrant youth is obvious in the Arctic region in order to maintain a sustainable demographic situation on the one hand, while on the other hand there is a need for further research on diverse issues relevant to immigrant youth in the Arctic, as there is less scholarly interest in that particular topic.

In Chapter 5, "Migrant integration in Finland: learning processes of immigrant women", Nafisa Yeasmin and Stefan Kirchner look at the integration of migrants in Finland, with a particular focus on the situation of immigrant women. Women often face particular challenges when it comes to integrating into host societies. Language skills and participation in the labour market are important for the integration of migrants. The latter in particular, and therefore also the former indirectly, can be challenging. This is especially so in the case of migrants from countries in which the dominant cultural attitude towards women is very different from the situation in the relatively egalitarian Nordic countries, such as Finland. Discriminatory attitudes, practices and even laws concerning women in their countries of origin can lead to limited access to education or to the ability to work outside the family home. Once in a Nordic host country, this discrimination experienced in their country of origin can create significant hurdles when they enter the labour market. Because of the importance of employment to the integration of migrants into host societies, the structural disadvantages suffered by women prior to migration continue to impact them in the new country. This can lead to specific learning needs among migrant women in the Nordic countries. In their research, Yeasmin and Kirchner outline migrant women's learning experiences in Finland. By placing their research in the wider theoretical context of migrant integration, they show how these learning processes, facilitated by the egalitarian nature of Nordic societies, can help immigrant women overcome such obstacles to integration. Immigrant women are a vulnerable group in the Arctic, and the integration of immigrant women is a threat to the host society in many regards. A women-specific integration approach can strengthen the resilience of immigrant women in the Arctic. The chapter addresses many gender-specific vulnerabilities that hinder immigrant women's integration and it also identifies women-specific integration stresses and shocks.

Chapter 6, written by Dr Waliul Hasanat, Nafisa Yeasmin and Timo Koivurova, deals with the impacts of practising various kinds of personal law within a single state. The main focus is on Muslim Sharia law followed by immigrant people in the Finnish context. The chapter explores the ongoing debates on practising a personal law other than that of the mainstream people, and it includes the key factors responsible for creating a malicious attitude towards the practitioners of Muslim family law, along with the main challenges faced by immigrant Muslims living in Finnish Lapland. It analyses whether legal pluralism, based on religion, may fit well in Finnish societies from theoretical perspectives, along with relevant case studies. Subsequently, the chapter tries to assess the major challenges of recognising non-state law from the viewpoint of an integral approach, as there are a few core needs of people that often vary from religion to religion, culture to culture and situation to situation. On the one hand, mainstream understanding of the needs of immigrants in their host country may lessen their struggle to settle in the new state, while, on the other hand, immigrants' respecting the family laws and cultures of the natives in public places may reduce locals' anxiety to protect their own culture.

In Chapter 7, Dr Juha Suoranta and Robert FitzSimmons focus on how the dramatic increase in the number of refugees entering Finland in 2015 and 2016 has led to challenges for society and the public administration in Finland, a country that, although hardly homogeneous, had seen little inward migration for a long time. Although the number of migrants reaching Finland pales in comparison with the number of refugees taken in by other countries, including by neighbouring Sweden, the increase in migrant numbers has led to numerous political discussions in Finland. In addition to the traumatic experience of having to leave their home countries, refugees also face challenges once they have arrived in a safe environment. Juha Suoranta and Robert FitzSimmons look at the experience of migrants who have come to Finland and who have applied for international protection there. In their chapter, entitled "Living in nowhere", they give a voice to those who have arrived in Finland and who are now awaiting a decision on their application for international protection. Although some applicants live in private accommodation, most asylum seekers in Finland live in reception centres. It is these asylum seekers that Suoranta and FitzSimmons talked with for their research. Their chapter looks at the fact that not only are reception centres for asylum seekers located in geographically remote locations (from the migrants' perspective), but also that asylum seekers may often feel disconnected from society so as to be living "nowhere" – neither in the society of the home country that they fled nor in Finland, a society about which they receive very little information. Prior to a decision on the application for international protection, asylum seekers are barely integrated into Finnish society. Often, the multiculturalism of living in a reception centre means that there is no shared language between asylum seekers living in a specific reception centre, which increases the feeling of isolation. Through their research, Suoranta and FitzSimmons have identified the needs of asylum seekers and simple possibilities for the improvement of their situation.

Chapter 8, written by Amelia Merhar, focuses on another important ethnic minority in the Circumpolar North, namely the indigenous communities and their socio-economic and cultural challenges. She investigates the situation of indigenous youth in Yukon (Canada) and, more precisely, the problem of moving from the child welfare system and foster care into adulthood. Unfortunately, as she stresses: "[the] child welfare system and its failings are especially devastating for indigenous communities, as First Nation, Inuit and Métis children under 14 make up 52.2 percent of all children in foster care" (Chapter 8, p. 000). Based on innovative, quality research, she seeks to analyse how the movement of youth within welfare system institutions and foster centres impacts their adult life and their understanding of home. Her study offers a set of important policy recommendations to make the welfare system more friendly towards indigenous youth – for instance, by making housing policies more flexible in terms of entry.

In Chapter 9, Stefan Kirchner and Susanna Pääkkölä address the question of protection from low temperatures from the perspective of human rights. Although the authorities in Northern Europe have been able to provide adequate housing for refugees even during the height of the refugee crisis, this has not always been the case in other European countries. In light of the particularly harsh climatic conditions and the limited predictability of migrant numbers, the chapter attempts an interdisciplinary look at the role of shelter and the right to live in healthy conditions. In many parts of the world, migration remains inherently dangerous, as is evidenced time and again by the loss of life due to the dangers of transportation – for example, during attempts to cross the Mediterranean Sea in small boats or to hide in the back of a truck. The years 2015–2016 saw a large number of migrants cross the border between the Russian Federation and Norway at the Storskog border crossing, many of them on bicycles. Most of the migrants had no prior experience of the harsh climate in the area, which is several hundred kilometres north of the Arctic Circle, and they were exposed to low temperatures. Low temperatures pose severe health risks, many of which are easily underestimated. This is especially so in the case of wind chill effects. Hypothermia – a dangerously low body temperature – is a particular threat to children and elderly people. Although the affluent countries of the European North provide adequate housing facilities for registered migrants, cold-related deaths of migrants have been reported in Eastern Europe. Therefore, there appears to be a need to clarify the human rights obligations of states in regard to the protection of migrants from the health threats posed by low temperatures.

After introducing the reader to the particular threats posed to human health by low temperatures, the chapter then provides an overview of the relevant international legal obligations that have been undertaken by Norway, Sweden and Finland. This overview includes not only general international human rights treaties, but also special treaties, such as the Convention on the Rights of the Child, the Convention on the Elimination of All Forms of Discrimination against Women and the Convention on the Rights of Persons with Disabilities. Particular emphasis is placed in the importance of the realisation of internationally defined human rights at the local level.

In his chapter on immigrant entrepreneurship, Chapter 10, Jan Brzozowski integrates the well-known theory of mixed embeddedness within the conceptual framework of community resilience. His chapter provides insights on how immigrant entrepreneurship could become a sustainable source of income for immigrants and their families, thereby contributing to their successful socio-economic insertion into the Arctic society. He demonstrates that immigrant entrepreneurship is not a process that can be left to immigrants alone in laissez-faire way. Just the opposite: public intervention is very welcome to provide solid guidance to immigrant entrepreneurs. Therefore, creating incentives for breaking-out strategies from vacancy-chain openings to more profitable sector markets that offer chances for firm development are needed. This implies building support structures for the entrepreneurial activities of newcomers, including community-based social enterprises, which could serve as "incubators" for future, real entrepreneurial activities. In this way, both the migrant communities and receiving regions would benefit, as sound support schemes for immigrant entrepreneurs would enable better usage and exploitation of their economic potential for the benefit of the entire society.

The book returns to issues concerning integration and education in Chapter 11, by Kirchner, on "Migration and Sustainable Development in the European Arctic". In this chapter, the author investigates the role of small, remote communities in the northernmost rural regions of Europe in the integration of migrants. It is shown that local communities are best suited to facilitate the integration of migrants, including refugees, but also that local communities require different types of support from higher-level authorities in order to be successful in these efforts. By linking integration to education and sustainable development to human rights, it is furthermore shown how migrants can contribute to the economic development of regions that otherwise might be restricted to non-sustainable forms of income – for example, through an excessive focus on tourism or extractive industries.

In Stefan Kirchner's chapter, we return to the impact that the 2015–2016 refugee crisis had on communities in Northern Europe and see that there are also positive effects of migration. Like rural regions elsewhere, many remote parts of the Arctic are suffering from demographic losses as young people leave the countryside in favour of opportunities for work or education in larger cities. This trend also affects the rural areas of the European High North, which is defined here as including Nordland, Tromsø and Finnmark in Norway, Norrbotten and Västerbotten in Sweden, and Lapland in Finland. Decreasing population numbers lead to a decrease in public services at the local level, such as the availability of village schools, which in turn makes rural regions even less attractive residential areas. Many remote Arctic and subarctic regions are also characterised by a limited choice of income opportunities for local residents. While environmental pollution and climate change lead to direct and indirect challenges to indigenous and traditional livelihoods, such as reindeer herding, the combination of increasing accessibility due to climate change and economic globalisation often leads to unsustainable forms of local economic development. National governments tend to emphasise unsustainable forms of development such as mining or hydrocarbon extraction, often for the economic benefit of outside corporations and with limited job opportunities for local residents. In many places in the European High North, tourism has not created the economic benefits that might have been hoped for locally, also due to the flow of income generated in the remote Arctic areas away from the region. Achieving sustainable economic development remains a challenge in many parts of the Arctic. As will be shown in Kirchner's chapter, migrants can play a role in supporting the economic development of small Arctic communities. Although migrants cannot be expected to fully reverse demographic trends that have been visible for many years, they can contribute to the economic development of remote Arctic regions and can create new opportunities for themselves and for longtime residents alike. In addition to the already positive political, legal and economic framework provided by the Nordic countries, this will require active integration by migrants. These experiences are then be placed in the context of the human rights dimension of sustainable development. It is shown how some small, remote communities in the European High North have become socially attractive new homes for migrants from different parts of the world, and how mutual benefits can be created together.

Concluding this chapter, we strongly believe that the aforementioned studies provide an interesting blend of papers that should serve as a point of reference for future studies on Circumpolar North migration and population challenges. Most of our studies have an explorative character and do not show ambitions to generalise over the entire Arctic region; however, there are obvious lessons and insights that each Arctic region could draw from these stories. Yet, we clearly see a need for more intensive, transnational scientific co-operation in Arctic research on socio-economic issues. Moreover, such co-operation should also be directed towards better data collection, which would enable quantitative studies to be conducted and provide population projections for local policymakers.

References

Berkes, F., & Ross, H. (2013). Community resilience: toward an integrated approach. *Society & Natural Resources*, 26(1), 5–20.

UArctic. (2017). UArctic Thematic Network on Arctic Migration. Available at www.uarctic. org/organization/thematic-networks/arctic-migration/

United Nations. (2015). *Population 2030: Demographic Challenges and Opportunities for Sustainable Development Planning*. New York: UN, Department of Economic and Social Affairs, Population Division.

Yeasmin, N. (2018). *The governance of immigration manifests itself in those who are being governed*. Doctoral dissertation. Lapland: University Press.

Part II

Youth perspective in the Arctic

2 The impact of superdiversity on the educational system

A mirror image of utopia or dystopia?

Nafisa Yeasmin and Satu Uusiautti

Introduction

According to the OECD reports of 2018, there are disparities in educational outcomes between foreign-born children and native-born: foreign-born children are more likely to underperform in mathematics (OECD, 2018; see also Kirjava- inen & Pulkkinen, 2017). Levels and Dronkers (2008) have found that immigrant students from western Europe, northern Africa, southern and central America, and western Asia performed worse than the autochthonous population. Indeed, many previous studies state that children and youth are at risk of confronting challenges in the educational systems in Europe (Gogolin, 2011). Many of these European countries are relatively new to receiving immigrants, in contrast to countries such as Singapore, Canada, and Great Britain. Diversity-related problems concern not only the educational sector, but also various other sectors. According to Steven Vertovec's superdiversity concept, it has been observed that superdiversity has a big impact on the educational sectors of Europe.

In this article, we discuss two countries that have performed well in the Programme for International Students Assessment (PISA) but have quite different educational systems and immigration structures. Finland, one of the top PISA- ranked countries, is also similarly facing the same challenges that have stemmed from the influx of immigrants in recent years. Finland, along with the rest of Europe, has been trying to recognise risk factors and has been identifying the reasons behind the increase in the dropout rate among immigrant youth (OECD, 2018). According to the same source, foreign-born children or children with foreign-born parents are at risk of self-censoring their aspirations.

According to PISA, insufficient adaptation causes school failure in Finland (Yeasmin & Uusiautti, 2018). Immigrant peers are considerably behind their non-migrant peers, although Finland is not yet listed as a superdiverse community compared with Singapore. Singapore, a superdiverse country, still scored top in the last PISA, whereas Finland, with a minimal number of immigrant children in schools, is facing several problems in integrating immigrant peers into schools. Our main research goal is to complete a comparative analysis with a focus on iden- tifying the integration dynamics that have an educational impact on immigrant children in (super)diverse school contexts; in addition, an impact assessment of the diverse classroom is the main emphasis of the research.

Our explorative research sought the answers through a comparative analysis between Singapore and Finland, two countries top-ranked by PISA (Yeasmin & Uusiautti, 2018). The education sectors of Finland and Singapore are the polar opposite of each other; as described previously, the Finnish education system is a mirror image of utopia, and the Singaporean educational system is highly reminiscent of the mirror image of utopia (Taffertshofer & Herrmann, 2007; Waldow, Takayama, & Sung, 2014). This discussion leads the debate on the educational systems of these two top-scoring PISA countries. We have noticed that ethnic diversity in Finland may worsen social interaction between majority and minority children in schools, and this delicate situation also requires special intercultural competences from the teachers (Yeasmin & Uusiautti, 2018; see also Taskinen, Uusiautti, & Määttä, 2019). In a superdiverse context, less interaction between children in diaspora has been observed, which also seems to create a competitive educational environment between the elite diaspora and the underprivileged diaspora. Overall, ethnic diversity can stimulate creativity among students, although it can also create diverse and hierarchical social classes (Vasu, 2012).

The diversity in the educational systems in Europe caused by the presence of immigrants is generating debate among politicians, locals, and society as well. A proportion of immigrants hinder the standard of native-born students' results, as discussed in previous research (Ammermüller, 2007). Previous research also states that the higher the diversity proportion in the school, the lower the quality of education (Thalhammer, Zucha, Enzenhofer, Salfinger, & Orgis, 2001). However, this assumption is not straightforward and should be viewed critically. It is not self-evident that cultural diversity affects the host country's educational system in either a positive or negative manner (Moreland, Levine, & Wingert, 1996). On the contrary, the Singaporean quality of education has reached a level of excellence even though the country is culturally superdiverse. This diversity-versus-superdiversity framework has influenced this study's theoretical foundation.

This chapter draws from the definitions of diversity and superdiversity and their possible impact on educational outcomes. The theoretical notions of diversity started from the demands of quality of life, innovation, demographic developments, and internationalisation awareness (Janssens & Steyaert, 2003). The concept of diversity actively accompanies the expanding heterogeneity of organisations (Janssens & Steyaert, 2003). Diversity has been explained as a moral and ethical perspective in some of the previous research (Antonio et al., 2004; Cox, 1995), whereas, in some other research, it has been discussed as immoral and unethical (Cox, 1991; De Dreu & Weingart, 2003). According to these perspectives, a diverse culture always places certain social groups in a disadvantaged position, which can lead to aggressiveness (Cox, 1991). The prevailing values and norms in a diverse society are hard to identify, and, therefore, organisational values could be immoral and unethical towards some groups. In a diverse system, a high tolerance level for ambiguity is needed (Cox, 1991).

The notion of superdiversity emerged in 2005 to explain the new diverse phenomenon of the 21st century (Vertovec, 2005, 2006, 2007). The increased movement of people causes a diverse composition from cultural, social, economic, political, and legal points of view, which changes the previous systems and patterns of host countries (Meissner & Vertovec, 2015). Understanding the advantages and disadvantages of diversity demands a new pattern of thinking about migration flows and their impact on host organisations (see also Uusiautti & Yeasmin, 2019). The practical aspects of diversity demand more attention from policymakers and practitioners to pluralism, which involves interconnected aspects of diversity and superdiversity. In our earlier study (Yeasmin & Uusiautti, 2018), we emphasised the multidimensional notion of the educational policies of two different countries in which both of the systems highlight certain kinds of configuration in the educational patterns of the respective countries.

Previous studies are yet to clearly identify exactly which particular situations can be categorised as diversity or superdiversity. To discuss these concepts, our purpose is to view the phenomenon of diversity through practical examples (see also Jakku-Sihvonen, Tissari, Ots, & Uusiautti, 2012). Therefore, in the chapter at hand, we have labelled Finland as a diverse country and Singapore as superdiverse. We base our definitions on the fact that ethnic groups still remain small in Finland (6 percent of the total population), whereas 38 percent of the total population of Singapore are citizens and the remaining 62 percent are something other than Singaporean (Singapore Population, 2018). The majority of the total population in Singapore is formed of ethnic groups. Keeping the majority–minority context at the forefront, along with both the number and size of the diverse groups (Crul, 2016), we can award Singapore the title of a superdiverse country.

Our special interest is in the educational systems of these two countries and the impact of superdiversity. By comparing these two educational systems, we pursue a discussion on the following questions: What makes an educational system superdiverse? What are the respective integration regimes that give the mirror images of utopia or dystopia? What are the superdiverse contexts that have an impact on the respective educational systems? What are those context-specific configurations of diversity that give us the mirror images of utopia and dystopia?

The Singaporean educational system as a utopian mirror image despite its superdiverse characteristics

Singapore is characterised as a global city (Olds & Yeung, 2004; Yingyue, 2012) in which education is a product of determination to make the city global. The country's good educational system gives a contextual backdrop impression to the world (Koh & Chong, 2014). The education system in Singapore is well established and well focused, which generates aspirations among students and their parents (Koh & Chong, 2014). The competitive education system in Singapore has an ideology manufacturing a sense of collectivity among groups and communities (Koh & Chong, 2014).

Multiple languages are spoken in Singapore, and this is the core constituent of diversity. Superdiversity has placed new demands on the Singaporean way of learning. For decades now, the government has been running various initiatives for learning and understanding the diversity in education (Koh & Chong, 2014).

By gathering information, brainstorming for decades, continuously changing policies, and adopting new policies in accordance with the present diverse situation, the end result is the success that Singapore reaps today (OECD, 2015). Language planning, such as making the mother tongue a compulsory subject for all students since 1966 and English as a first language, offers a logical approach and concept to assimilating cultural diversity (Bulle, 2011). These policies indeed showed respect and shared values towards the diverse community. This has created an assimilating learning style for immigrant families who are either Singaporean residents with permanent status or residents with temporary status.

Policies regarding native languages share neutral language policies and converging learning opportunities by accommodating the emotional well-being of ethnic groups logically. It opens plentiful opportunities to practise cultural traditions and strengthens racial identities. This opportunity offered by the native language policy gives both second-generation students and students who were born into ethnic Singaporean families a chance to study their native language, which also helps them to bond with transnational families. This opportunity gives a sense of creativity and helps to build social bonds and contacts (Tashakkori & Teddlie, 2003).

The national curriculum, which is controlled by the state, supports the academic performances of students in Singapore (Bulle, 2011; Waldow et al., 2014). This national curriculum also governs a nationwide meritocratic examination that determines students' academic performances. There are diverse examination processes for the selection of students into high schools and universities. The traditional pedagogical practices of examinations aid their competency paths in the PISA tests as well (Deng & Gopinathan, 2016).

The school excellence model (School Excellence Model, n.d.) is a particular initiative to create a close relationship between schools and the Ministry of Education. This model helps to maintain school organisation, the professional competences of community school teachers, and teachers' relationships with the parents and community (Ng & Chan, 2008). School leaders work with respective community leaders to improve the overall framework of schools, which is considered necessary for educational achievement (Ng & Chan, 2008).

In this utopia, communities, pupils, and teachers recognise and empower one another (Ng & Chan, 2008). The school excellence model is indeed a quality framework for schools, a model in which school agents can self-assess their quality; external inspections also take place to measure quality. This system empowers diverse community schools and develops school leadership experiences along with ensuring the quality of the community schools. These systems create accountability in schools to the authority (Ng & Chan, 2008). This model is a framework for school appraisal, and it follows a scoring system. Under the scoring system, each school can collect grades on attainment levels (school excellence model). This practice of awarding schools a grade improves the pedagogical culture, quality, and professionalism of the teachers in both community and public schools (Dimmock & Goh, 2011).

The direct, official policies for understanding diversity and the announcement of this document publicly state that diversity is essential for sustaining the population and economic diversity of Singapore. Singapore may have diverse geographical and ethnic backgrounds, but all Singaporeans seem to share a certain set of key values and aspirations (Population White Paper [PWP], 2013). It states that all systems are fair and respect one another's culture (PWP, 2013). The government's open and transparent policy encourages acceptance of superdiversity, which encourages people with different ethnic backgrounds to stay in Singapore (PWP, 2013).

A special feature of the Singaporean model is that upgrading policies are based on practical experiences and negotiations of diversity. Perspectives and policies that support diversity are openly encouraged.

> The corporatist model of governance and multiculturalism in Singapore may come to an end owing to the development of greater inter-communal cohesion among Singaporeans – where the state is exclusively responsible for legitimating and enfranchising group and individual participation in public affairs, while also being responsible for managing the harmonious interaction among these groups and individuals.
>
> (Vasu, 2012, p. 736)

An immutable understanding of intergroup differences inhibits segregation in the classroom. Pupils of similar race and socio-economic status are used to going to the same school. Thus, this also inhibits bullying in classrooms with homogeneous characteristics. Educational competition raises awareness and generates aspirations to succeed among pupils and parents (Brown, 2012; Francis & Wong, 2013). On the other hand, the social studies curriculum in schools includes social cohesion within a diverse society (Singapore Ministry of Education, 2008; see also Ho, 2009). The multicultural education initiative also incorporates empathy towards and tolerance of others' ethnicity, beliefs, and religion, which in turn inspires discussion on utopian thinking.

The considerate Finnish education system as a utopia despite its less diverse nature

Some researchers (e.g., Taffertshofer & Herrmann, 2007; Waldow et al., 2014) have described the Finnish education system as the image of an educational utopia for learners. It has become the model for world education (Sahlberg, 2010; Waldow et al., 2014), and its quality and equality produce a utopian image. The Finnish education system is based on a regime that declares it a public service and free of charge.

> Basic education, upper secondary education, and vocational education are financed by the state and local authorities (municipalities). General education and vocational education are provided by local authorities, while universities are autonomous and financed by the government.
>
> (Yeasmin & Uusiautti, 2018, pp. 207–237)

Ensuring basic education for all children, regardless of their background, is explicitly mentioned in the integration policy (Ministerial Integration Group, 2016). All children, be they residents with permanent status or those with temporary status, are obliged to have preschool education. Even the children of asylum seekers are awarded opportunities for free education. All students, regardless of their ethnic backgrounds, are entitled to receive free vocational education and training instruction, free meals every school day, and free accommodation assigned by the respective educational institutions (European Agency, 2018). Parents do not have to bear any additional expenses for their children's education. There is no pressure for pupils to take after-school private tuition in order to achieve the highest possible level of performance. The Finnish education system is rarely competitive; thus, children have plenty of time to spend on after-school activities, which is very much a relaxed attitude and atmosphere compared with the Asian systems (Yeasmin & Uusiautti, 2018).

In recent years, schools in Finland have been focusing on the appropriate implementation of inclusive education to teach all levels of students in the same classroom (e.g., Lakkala et al., 2016; see also Taskinen et al., 2019). Inclusive learning is a recommended method of learning in Finnish schools. Kolb's accommodating learning style (Kolb & Fry, 1974) relies on intuition to learn from other's experiences (McLeod, 2017). In this method, immigrant children are in the same classes as Finnish children, which allows them to learn through teamwork and by exchanging experiences (see Taskinen, 2017). Recent research in education has looked at ways of encouraging inclusive education in schools. The Finnish education system is decentralised, and all municipalities exercise a wide range of autonomy to design a local curriculum based on the core national curriculum. Local curricula are influenced by local educational needs (Yeasmin & Uusiautti, 2018). Facilitating the integration of immigrant pupils, handling human rights in schools, the rights of children, equity, and equality are all part of the Finnish national immigration policy, and the curriculum indeed includes understanding of diverse cultures and religious traditions (Opetushallitus, 2010).

The Finnish government has employed a positive discrimination method by increasing funds and integration support for the education of immigrant students and their teachers (Yle, 2017). One of the policies is to provide more resources for teachers' continuing education to increase their intercultural competences, as previous research stated (Yeasmin & Uusiautti, 2018).

> [Finland]gives schools extra funds if they are situated in relatively poor areas or have a disproportionately high number of children with special needs. It tops up these funds with €1,000 (£875) a year for each child on the school's roll who has lived in Finland for less than four years.
>
> (Shepherd, 2011, n.p.)

Immigrant children need special education to acquire Finnish language skills when they first arrive in Finland.

Finnish teachers have plenty of freedom in how they design their lessons and what kinds of teaching method they use. School days are relatively short, and there is not much testing of students in basic education or supervision of teachers' performance in Finnish schools (e.g., Baines, 2007; Gamerman, 2008). Among others, these features of the Finnish education system have puzzled researchers and they have, in part, been recognised as one of the reasons for the Finnish students' generally high scores in PISA comparisons (e.g., Simola, 2005; see also Paksuniemi, Uusiautti, & Määttä, 2013). Finnish teachers are sceptical about giving students frequent standardised tests and are not accountable for their professional autonomy (Bastos, 2017).

Superdiversity as a mirror image of dystopia in the Singaporean education system

In superdiverse systems, reform and creativity are a constant must. The growing marketisation of Singaporean education also creates stress and competition among educational agents. Somehow, this competition creates force accountability in schools (Tan & Gopinathan, 2000). The state takes centre stage in Singapore, and the state is responsible for social cohesion and managing the harmonious interaction between groups and individuals (Vasu, 2012). This corporatism tendency of Singapore has created strong challenges in the school governance systems. The bilingual language policy in Singapore has been established in order to respect superdiversity, and the policy has created a dystopian discussion of segregation. The major ethnic groups of Chinese, Malay, and Tamil use various native languages and dialects alongside English, which disrupts the legitimacy of the English language. It also makes it difficult for the government to ensure the learning of standard English (Tan & Ng, 2011).

In an English-speaking multilingual society, forced bilingualism can create obstacles to learning perfect English, as speaking English brings privilege and social prestige in Singapore (Ng, 2014). Studying native language in schools requires a realistic level of thinking, as it creates a language load for students in the examinations when they leave primary school. Students are learning languages for the purpose of passing the exam, not with the goal of real, functional language knowledge (cf., Taskinen, 2017). This creates more pressure on top of the existing pressure for high achievement in schools, where speaking English is official and a sign of the elite.

This policy categorises individuals (Rampton, 1990, 2007) and creates schooling diaspora. Diaspora strategies are complicated and controversial interpretations of integration. The Singaporean educational strategies encourage immigrants to live in their own diaspora and create diasporic networks. Diversity conflicts end in diasporic conflict in multiracial nationals in Singapore. Ethnic diaspora are struggling with their national identities. As Singaporeans are always dependent on government decisions, the government is also striving to construct national ideologies and close the diasporic relational gap (Mishra, 1996). Not only does every ethnic group create its own diaspora within larger diaspora and even elite diaspora for those elite Indians, Chinese, and Malay, but cultural diaspora also leads to schooling diaspora and working diaspora. White-collar ethnic groups reside in elite diaspora and send their children to elite schools with high tuition fees (Subramanian, 2006). New arrivals of ethnic groups form new racial politics. Superdiversity is, therefore, creating socially diverse classes between and beyond ethnic groups (Subramanian, 2006).

On the other hand, skilled workers or professional talents from abroad (e.g., India or China) are always desired by the state, but the government's treatment of them and the elite groups is not consistent (Subramanian, 2006). Obtaining resident permits is easier for them than for blue-collar workers (Subramanian, 2006). The system also prioritises its own residents and sometime even prioritises admission for its own residents (The Expat, 2015), which then creates segregation. The main target is reaching a high-quality standard of living in Singapore and making an excellent impression on the rest of the world, which also creates hierarchy between and within the groups. These diasporas also support advantage and disadvantage schools (Subramanian, 2006; Teng, 2018). Social statutes depend on educational status and economic survival. According to an OECD report, "Nearly half of low-income students in Singapore are concentrated in the same schools" (Teng, 2018). Children's education very much depends on the wishes and opportunities of the parents and also how much they can afford to pay for their children's education.

Excessive tuition fees to improve children's academic achievement is a stressful issue for parents with a weak socio-economic status. Improving skills and performance levels in order to belong to a better class, both in the diaspora and in overall society, creates pressure among parents and students and is seen as a dystopian image. After-school tuition centres also differ between the diverse races (Subramanian, 2006).

Social classes (Gardiner, 2015) are a stress factor for many parents and children. No matter which social class they belong to, their goal is always to climb higher on the social ladder. Additional pressure to perform at a high level comes from the desire to win admission to the best school, college, or university.

The wider detrimental effect of this educational inequality needs further analysis. We focus on inclusive education in Europe, particularly in Finnish cases, which contradict this dystopian thinking.

A less diverse education system as the Finnish dystopia

Many outsiders view Finland as a monoculture, as the percentage of immigrants is still low (Shepherd, 2011). According to an article in *The Guardian*, "[a] growing number of Finns are said to be removing their children from ethnically diverse primary schools, and some are reported to be demanding a cap on the number of non-Finns in a classroom". Finnish children's attitudes towards immigrant pupils also differ. The parents' socio-economic status has little effect on children's educational attainment, but gaining knowledge through non-formal activities or extracurricular activities depends greatly on the social status of the parents (Ong & Yeoh, 2012). A different level of social status is related to various levels of self-esteem, aptitudes, and relatedness (Cox, Duncheon, & McDavid, 2013). In accordance with our previous research (Yeasmin & Uusiautti, 2018, pp. 207–237):

> immigrant students would need to be evaluated in relation to their participation in the classroom, integration in the society, socialization, and development of language skills. If the immigrant student becomes accepted in the group and finds friends, the chances of understanding the school culture in the host country are better and then succeeding well also academically will not only be more likely but also easier.

Proper integration measures are needed, as higher school dropout rates are present among immigrant children from outside the EU after basic compulsory education (Kilpi-Jakonen, 2011). The performance level of those students is poorer than that of their counterparts (OECD, 2018). Immigrant children seem to lack educational well-being and would benefit from special support towards better integration into Finland (Matikka, Luopa, Kivimäki, Jokela, & Paananen, 2013). They lack friends, feel inferior both inside and outside school, and make an easy target for bullying (Matikka et al., 2013).

Immigrants from less-developed countries perform below standard when compared with the non-immigrant population (Kilpi-Jakonen, 2011). Despite several immigration policies and the positive discrimination policies of the government, immigrants still find themselves an isolated group in Finnish society (Yeasmin & Uusiautti, 2018), which means that governmental attitudes towards immigrants are positive, but immigrant families tend to be excluded from society, leading to a division of social classes. Perhaps the educational background of immigrant parents also has an impact on their socio-economic status in Finnish society (Kilpi-Jakonen, 2011; Matikka et al., 2013). Their attitudes to and their willingness to adopt the Finnish lifestyle can be seen as one determinant of integration as well (see also Valtonen, 2016). The lower employment rate among foreign-born people in Finland indeed affects their social status (OECD, 2018). Immigrants' skin colour and ethnic origin can place them at a disadvantage. Finnish workers are clearly less comfortable working with immigrants, but, on the other hand, awareness of discrimination is high in Finland (OECD, 2018).

Immigrant parents with low social status seem to be less interested in being involved with their children's schooling systems (Heath & Kilpi-Jakonen, 2012). Lastikka and Lipponen (2016) found that immigrant parents are not aware of the Finnish education system, and actions to foster dialogue and mutual understanding and to encourage co-operative partnerships are needed, in addition to the support and individualised attention that immigrant families receive.

In countries with less diversity or with a smaller population of immigrants, immigrant children in particular are vulnerable, and their performance level in PISA is lower that of their native-born counterparts; Finland is one of these countries (OECD, 2015). In the Finnish education system, immigrant children have to learn the Finnish language. It is considered vitally important that they learn Finnish as their second language.

No doubt, immigrant children can gain sound structural knowledge of Finnish. However, as Finnish is their second language, they may not be able to express all their feelings and may have difficulties in understanding mathematics or other school subjects (OECD, 2018). Children's cognitive development can be hampered if they are discriminated against or rejected by their friends because of their language ability or inability to communicate because of their language problems (Lasonen, 1978; Vuorenkoski, 2000), which may lead to depression among students and can increase the dropout rate among immigrant children (Heikkilä & Yeasmin, 2017). Sometimes, language barriers hinder their inclusive learning opportunities in mainstream classes.

On the other hand, the Finnish education system, with its low rate of formal examinations, does not create a competitive environment of learning and test-based accountability. Examinations can facilitate the mapping of students' level and capacity in various subjects, which indeed supports students to develop their pedagogical needs, and similarly teachers can also gain knowledge of and insight into the students' strengths and weaknesses in different subjects and support them accordingly (Leskisenoja & Uusiautti, 2019; Nilsson & Bunar, 2015). Teachers can redesign their teaching methods according to the test results and give higher priority to the particular subjects in which the children are weaker, which overall is not a negative action (Bastos, 2017). But, in Finland, teachers can focus on teaching without needing to subject the pupils to frequent tests (Sahlberg, 2010): the only national examination for students is the matriculation examination at the end of upper-secondary education in Finland. Despite the low rate of competition in Finnish schools, Finnish students constantly report low levels of happiness due to a sense of not being cared for by teachers at school (e.g., Harinen & Halme, 2012): Finnish students' well-being at school scored the fifth lowest among 65 countries within this category (OECD, 2016). At the moment, the well-known, lauded, research-based teacher training in Finland (see Uusiautti & Määttä, 2013) will have to improve in the field of caring about students (see Leskisenoja & Uusiautti, 2019).

Utopia and dystopia: diversity and superdiversity

Managing a culturally diverse society is not an easy task for social institutions. There are positive and negative consequences of diversity and superdiversity, which require a suitable conceptual framework for managing diversity. Though challenges exist in more or less every society, every society needs a reshaping of its regular, monotonous life conditions through diversity. The transformation of a society is a process of some inclusion and, concurrently, some exclusion among social groups (Castells, 1996). But the end of this reshaping leads to a new social contract. Some groups are valued, and some are devalued in such a diverse society. There are both costs of and benefits to ethnic diversity (Yeasmin, 2018). Superdiversity is a complex phenomenon that inspires debates on how to understand the realities and consequences of diversity. As we see from the above discussion, when it comes to a utopia, the impetus towards hopes for a positive focus on a diverse education system is readily identified. The reasoning behind these countries' high PISA results demonstrates a hope for a utopian future in a diverse/superdiverse society. At the opposite end of the spectrum, if we focus on social justice, then this discussion demonstrates a sense of frustration and fear of a dystopian future. Access to good education appears as a concern and a stressful phenomenon in Singapore, whereas access to quality education is easier in Finland. The impact of similar characteristics creates both fear and hope, depending on the social structures. Under some circumstances, diversity provides a perfect image of a perfect society, as diversity improves some educational aspects – resembling a utopia in a perfect way. Therefore, the cognitive value of the consequences and warnings of diversity or superdiversity to the educational system highlights that achieving the unachievable is a process of striving. In regards to the utopian discussion above, it suggests a need for a dialogue on diversity and superdiversity to generate discursive argument on dystopian aspects in a more skilful way.

We cannot step back from the dystopian impacts of diversity or superdiversity. We can hope for good solutions from our fear images that can lead to an understanding of the utopian constellation. Managing diverse classrooms is challenging, but a diverse, bilingual or monolingual class may also have higher levels of divergent thinking along with cognitive flexibility (Cox, 1991).

We can conclude from our analysis that, if there is a problem, there are also solutions. Diversity and superdiversity may be viewed as a route to dystopia. By changing the perspective, they can simultaneously appear to be the makings of a utopia. The analysis is to show that we need to be ready to acknowledge a variety of perspectives in order to recognise those features of education that are functional and those that are not.

Conclusion

The ultimate challenge of diversity or superdiversity in education systems is the rise of pluralism in the educational institutes. Also contributing to the diversity of classrooms are the numerous subcultures, different identities, and different choices of norms (Colombo, 2013). Shaping this heterogeneity requires co-responsibilities to understand diverse perspectives, which include critical thinking, adaptation, and ethical consideration in diversity management. All pluralistic views in a diverse community represent a need for innovative and effective learning to confront new situations. New attitudes towards new social structures and a new set of values could create a mirror image of utopia or dystopia in the respective society. A (super)diverse community requires key abilities to understand the concept of pluralistic approaches. If a society has the capacity to understand this and can utilise the opportunities of diversity, it can be called an example of utopia (Cox, 1991; Hoffman, 1959; McLeod & Lobel, 1992; Milliken & Martins, 1996; Northcraft, Polzer, Neale, & Kramer, 1995; Thomas & Ely, 1996; Watson, Kumar, & Michaelson, 1993). However, if the society loses the capacity to learn about diversity, it can create the negative result of failure

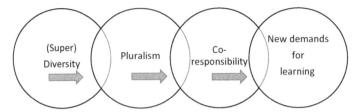

Figure 2.1 Impacts of superdiversity (designed by NY)

(Cox, 1991; Ibarra, 1992, 1995; Watson et al., 1993), which could symbolise the dark mirror image of dystopia (Claisse & Delvenne, 2015). Figure 2.1 illustrates the aforementioned processes.

We know that a utopian way of thinking in any given social structure outlines relative improvement through learning and making accommodations to the diverse arrangement. Converging and assimilating socio-ontogenetic capacities in order to absorb diverse preferences are usual traditions of utopian (Spiro, 1956) thinking (Dahrendorf, 1958; Kateb, 1971). Both of these societies are dreaming about a better education system in a society, even if it is (super)diverse. In Finland, we are making predictions about our future education system, in what is becoming an increasingly diverse community. In that case, we can rely on a critical observation of other diverse countries and their education system, such as Singapore. If Finland also continues to grow in diversity owing to increased immigration, similar to Singapore, can Finland maintain its educational quality in the future? Here, the dystopian posture has been used to defend the notion of diversity in education. When it comes to diversity and whether it is super or not, Singapore proved its success in the last PISA test, as well as in its overall educational system (Yeasmin & Uusiautti, 2018).

Thus, critical reflection on solutions that have already been implemented in diverse and diversifying situations is crucial. We have to be able to open our eyes to various scenarios. Finland cannot rest on its laurels with its eternal happiness and success; it must face reality and the new demands for quality education that the diversifying population creates. Neither can Singapore be complacent – it has to evaluate the purposefulness of its system, too. However, perhaps the comparison of the utopian and dystopian mirror images of these two fundamentally different countries, with their unique education systems, can provide a fruitful ground for further discussion.

References

Ammermüller, A. (2007). Poor background or low returns? Why immigrant students in Germany perform so poorly in the Program for International Student Assessment. *Education Economics*, 15, 215–230.

Antonio, A. L., Chang, M. J., Hakuta, K., Kenny, D. A., Levin, S., & Milem, J. F. (2004). Effects of racial diversity on complex thinking in college students. *Psychological Science*, 15, 507–510.

Baines, L. (2007, October). Learning from the world: Achieving more by doing less. *Phi Delta Kappan, 89*(2), 98–100.

Bastos, R. M. B. (2017). The surprising success of the Finnish educational system in a global scenario of commodified education. Retrieved from www.scielo.br/scielo.php?pid=S1413-24782017000300802&script=sci_arttext&tlng=en

Brown, G. (2012). The place of aspiration in UK widening participation policy: moving up or moving beyond? In J. Horton, P. Kraftl, & F. Tucker (Eds.), *Critical Geographies of Children and Youth* (pp. 97–113). Bristol, UK: Policy Press.

Bulle, N. (2011). Comparing OECD educational models through the prism of PISA. *Comparative Education, 47*(4), 503–521.

Castells, M. (1996, 22nd ed. 2009). *The Rise of the Network Society, the Information Age: Economy, Society and Culture* (Vol. I). Malden, MA: Blackwell.

Claisse, F., & Delvenne, P. (2015). Building on anticipation: Dystopia as empowerment. *Current Sociology, 63*(2), 155–169.

Colombo, M. (2013). Pluralism in education and implications for analysis. *Italian Journal of Sociology of Education, 5*(2), 1–16.

Cox, A., Duncheon, N., & McDavid, L. (2013). Peers and teachers as sources of relatedness perceptions, motivation and affective responses in physical education. *Research Quarterly for Exercise and Sport, 80*(4), 765–773.

Cox, T. (1991). The multicultural organization. *Academy of Management Executive, 5*(2), 34–47.

Cox, T. (1995). The complexity of diversity: challenges and directions for future research. In S. E. Jackson & M. N. Ruderman (Eds.), *Diversity in Work Teams* (pp. 235–246). Washington, DC: American Psychological Association.

Crul, M. (2016). Super-diversity vs. assimilation: How complex diversity in majority–minority cities challenges the assumptions of assimilation. *Journal of Ethnic and Migration Studies, 42*(1), 54–68.

Dahrendorf, R. (1958). Out of utopia, toward a reorientation of sociological analysis. *American Journal of Sociology, 64*(2), 115–127.

De Dreu, C. K. W., & Weingart, L. R. (2003). Task versus relationship conflict, team performance, and team member satisfaction: A meta-analysis. *Journal of Applied Psychology, 88*, 741–749.

Deng, Z., & Gopinathan, S. (2016). PISA and high-performing education systems: explaining Singapore's education success. *Comparative Education, 52*(4), 449–472.

Dimmock, C., & Goh, J. W. P. (2011). Transformative pedagogy, leadership and school organisation for the 21st century knowledge-based economy: the case of Singapore. *School Leadership and Management, 31*(3), 215–234.

European Agency. (2018). Country information for Finland – financing of inclusive education systems. Retrieved from www.european-agency.org/country-information/finland/financing-of-inclusive-education-systems

The Expat. (2015). Retrieved from www.cadogantate.com/en/moving-services/news/pros-and-cons-moving-to-singapore-children (August 30 2018).

Francis, B., & Wong, B. (2013). What is preventing social mobility? A review of the evidence. Retrieved from www.researchgate.net/publication/272164028_What_is_preventing_social_mobility_A_review_of_the_evidence (December, 2019).

Gamerman, E. (2008). What makes Finnish kids so smart? *The Wall Street Journal*. Retrieved from www.wsj.com/articles/SB120425355065601997 (July 28, 2017).

Gardiner, J. B. (2015). Social class and educational achievement in modern England: the impact of aspirations, attitudes, self-belief and cultural identity on working class pupil's educational achievement – a review of the literature. Retrieved from www.researchgate.net/publication/282003754_Social_Class_and_Educational_Achievement_in_Modern_

England_The_Impact_of_Aspirations_Attitudes_Self-Belief_and_Cultural_Identity_on_Working_Class_Pupil's_Educational_Achievement_-_A_Review_of_the_Lite (May 30, 2019).

Gogolin, I. (2011). The challenges of superdiversity for education in Europe. *Education Inquiry, 2*(2), 239–249.

Harinen, P., & Halme, J. (2012). *Hyvä, paha koulu. Kouluhyvinvointia hakemassa* [Good, bad school. Looking for well-being at school]. Helsinki: UNICEF Finland. Retrieved from www.nuorisotutkimusseura.fi/images/julkaisuja/Hyva_paha_koulu.pdf

Heath, A., & Kilpi-Jakonen, E. (2012). Immigrant children's age at arrival and assessment results. *OECD Education Working Papers, No. 75.*

Heikkilä, E., & Yeasmin, N. (2017). Haasteena saavuttaa onnistunut koulupolku [The challenge is to reach a successful school path]. In M. Körkkö, M. Paksuniemi, S. Niemisalo, & R. Rahko-Ravantti (Eds.), *Opintie sujuvaksi Lapissa [Making schoolpaths smooth in Lapland].* (pp. 133–147). Turku, Finland: Institute of Migration.

Ho, L. C. (2009). Global multicultural citizenship education: a Singapore experience. *The Social Studies, 100*(6), 285e293.

Hoffman, L. R. (1959) Homogeneity of member personality and its effect on group problem-solving. *Journal of Abnormal and Social Psychology, 58*, 27–32.

Ibarra, H. (1992). Homophily and differential returns: sex differences in network structure and access in an advertising firm. *Administrative Science Quarterly, 37*, 422–447.

Ibarra, H. (1995). Race, opportunity and diversity of social circles in managerial networks. *Academy of Management Journal, 38*(3), 673–703.

Jakku-Sihvonen, R., Tissari, V., Ots, A., & Uusiautti, S. (2012). Teacher education curricula after the Bologna Process – a comparative analysis of written curricula in Finland and Estonia. *Scandinavian Journal of Educational Research, 56*(3), 261–275. doi:10.1080/00313831.2011.581687

Janssens, M., & Steyaert, C. (2003). Theories of diversity within organisation studies: debates and future trajectories. Retrieved from https://core.ac.uk/download/pdf/6264654.pdf

Kateb, G. (Ed.) (1971). *Utopia*. New York: Routledge.

Kilpi-Jakonen, E. (2011). Continuation to upper secondary education in Finland: children of immigrants and the majority compared. *Acta Sociologica, 54*(1), 77–106.

Kirjavainen, T., & Pulkkinen, J. (2017). Miten lähtömaa on yhteydessä maahanmuuttajaoppilaiden osaamiseen. Oppilaiden osaamiserot PISA 2012 –tutkimuksessa [What is the connection between the country of origin and immigrant students' performance. Differences in student performance in PISA 2012 research]. *Yhteiskuntapolitiikka, 82*(4), 430–439.

Koh, A., & Chong, T. (2014). Education in the global city: The manufacturing of education in Singapore. *Discourse: Studies in the Cultural Politics of Education, 35*(5), 625–636.

Kolb, D. A., & Fry, R. E. (1974). *Toward an Applied Theory of Experiential Learning*. Cambridge: MIT Alfred P. Sloan School of Management.

Lakkala, S., Uusiautti, S., & Määttä, K. (2016), How to make the neighbourhood school a school for all? *Journal of Research in Special Educational Needs, 16*(1), 46–56. doi:10.1111/1471-3802.12055

Lasonen, K. (1978). *Research report. Jyväskylä,* Finland: Jyväskylä, Finland: Department of Education, University of Jyväskylä.

Lastikka, A.-L., & Lipponen, L. (2016). Immigrant parents' perspectives on early childhood education and care practices in the Finnish multicultural context. *International Journal of Multicultural Education, 18*(3), 75–94.

Leskisenoja, E., & Uusiautti, S. (2019). Human strength-spotting at school as the future foundation of "us" in the Arctic. In S. Uusiautti & N. Yeasmin (Eds.), *Human Migration in the Arctic – The Past, Present, and Future* (pp. 239–261). Singapore: Palgrave Macmillan.

Levels, M., & Dronkers, J. (2008). Educational performance of native and immigrant children from various countries of origin. *Ethnic and Racial Studies, 31*(8), 1404–1425.

Matikka, A., Luopa, P., Kivimäki, H., Jokela, J., & Paananen, R. (2013). *Kouluterveyskysely – Maahanmuuttajataustaisten 8. ja 9. Luokkalaisten hyvinvointi.* Terveyden ja Hyvinvoinnin Laitos (THL). Retrieved from www.julkari.fi/bitstream/handle/10024/116720/URN_ISBN_978-952-302-297-3.pdf?sequence=1&isAllowed=y (December 25, 2018).

McLeod, P. L., & Lobel, S. A. (1992). The effects of ethnic diversity on idea generation in small groups. *Academy of Management Best Paper Proceedings, N/A*: 227–231.

McLeod, S. (2017). Kolb-learning styles and experimental learning cycle. Retrieved from www.simplypsychology.org/learning-kolb.html

Meissner, F., & Vertovec, S. (2015). Comparing super-diversity. *Ethnic and Racial Studies, 38*(4), 541–555.

Milliken, F. J., & Martins, L. L. (1996). Searching for common threads: understanding the multiple effects of diversity in organizational groups. *Academy of Management Review, 21*(2), 402–433.

Ministerial Integration Group. (2016). Government Integration Programme for 2016–2019 and Government Resolution on a Government Integration Programme. A publication of the Ministry of Economic Affairs and Employment. Retrieved May 26, 2018, from https://julkaisut.valtioneuvosto.fi/bitstream/handle/10024/79156/TEMjul_47_2016_verkko.pdf?sequence=1

Mishra, V. (1996). The diasporic imaginary: theorizing the Indian diaspora. *Textual Practice, 10*(3), 421–447.

Moreland, R. L., Levine, J. M., & Wingert, M. L. (1996). Creating the ideal group: composition effects at work. In E. H. Witte & J. H. Davis (Eds.), *Understanding Group Behavior: Small Group Processes and Interpersonal Relations* (pp. 11–35). Mahwah, NJ: Erlbaum.

Ng, C. L. P. (2014). Mother tongue education in Singapore: concerns, issues and controversies. *Current Issues in Language Planning, 15*(4), 361–375.

Ng, P. T., & Chan, D. (2008). A comparative study of Singapore's school excellence model with Hong Kong's school-based management. *International Journal of Educational Management, 22*(6), 488–505.

Nilsson, J., & Bunar, N. (2015). Educational responses to newly arrived students in Sweden: understanding the structure and influence of post-migration ecology. *Scandinavian Journal of Education Research, 60*(4), 1–18.

Northcraft, G. B., Polzer, J. T., Neale, M. A., & Kramer, R. M. (1995). Diversity, social identity and performance: emergent social dynamics in cross-functional teams. In S. E. Jackson & M. N. Ruderman (Eds.), *Diversity in Work Teams* (pp. 69–96). Washington, DC: American Psychological Association.

OECD. (2015). PISA 2015 results in focus. Retrieved from www.oecd.org/pisa/pisa-2015-results-in-focus.pdf

OECD. (2016). *PISA 2015 Assessment and Analytical Framework*. Paris: Author.

OECD. (2018). *Working Together: Skills and Labour Market Integration of Immigrants and Their Children in Finland*. doi:10.1787/9789264305250-en

Olds, K., & Yeung, H. (2004). Pathways to global city formation: a view from the developmental city-state of Singapore. *Review of International Political Economy, 11*(3), 489–521.

Ong, F. C. M., & Yeoh, B. S. A. (2012). The place of migrant workers in Singapore: Between state multiracialism and everyday (un)cosmopolitanisms. In A. E. Lai, B. S. A. Yeoh, & F. C. Leo (Eds.), *Migration and Diversity in Asian Contexts* (pp. 83–106). Singapore: ISEAS-Yusof Ishak Institute.

Opetushallitus. (2010). Esiopetuksen opetussuunnitelman perusteet 2010. Määräykset ja ohjeet 2010:27 Helsinki, Finland: Author.

Paksuniemi, M., Uusiautti, S., & Määttä, K. 2013. *What Are Finnish Teachers Made of? A Glance at Teacher Education in Finland Yesterday and Today*. New York: Nova.

Population White Paper (PWP). (2013). *A Sustainable Population for a Dynamic Singapore*. Singapore: National Population and Talent Division. Retrieved from www.strategygroup. gov.sg/media-centre/population-white-paper-a-sustainable-population-for-a-dynamic-singapore (April, 7, 2020).

Rampton, B. (1990). Displacing the "native speaker": expertise, affiliation and inheritance. *ELT Journal, 44,* 97–101.

Rampton, B. (2007). Neo-Hymesian linguistic ethnography in the United Kingdom. *Journal of Sociolinguistics, 11*(5): 584–607.

Sahlberg, P. (2010). Educational change in Finland. In A. Hargreaves, A. Lieberman, M. Fullan, & D. Hopkins (Eds.), *Second International Handbook of Educational Change* (pp. 323–348). Dordrecht, Netherlands: Springer.

School Excellence Model. (n.d.) Retrieved from https://sites.google.com/site/consultingone/schoolexcellencemodel

Shepherd, J. (2011, November 21). Immigrant children benefit from education. *The Guardian*. Retrieved from www.theguardian.com/education/2011/nov/21/finland-education-immigrant-children

Simola, H. (2005). The Finnish miracle of PISA: historical and sociological remarks on teaching and teacher education. *Comparative Education, 41*(4), 455–470. doi: 10.1080/03050060500317810

Singapore Ministry of Education. (2008). *Combined humanities ordinary level social studies syllabus (Syllabus 2192) [electronic version]*. Retrieved from www.seab.gov.sg/SEAB/oLevel/syllabus/2008_GCE_O_Level_Syllabuses/2192_2008.pdf (no longer available).

Singapore Population. (2018) Retrieved from http://populationof2018.com/singapore-population-2018.html

Spiro, M. E. (1956). *Kibbutz; Venture in Utopia*. Cambridge, MA: Harvard University Press.

Subramanian, A. (2006). From colonial segregation to postcolonial "integration" – constructing ethnic difference through Singapore's Little India and the Singapore "Indian". Retrieved from https://core.ac.uk/download/pdf/35461221.pdf

Taffertshofer, B., & Herrmann, G. (2007, June 18). Paradise in the North. More teachers, no grade pressure, no repetition of grades. *Süddeutsche Zeitung*.

Tan, C., & Ng, P. T. (2011). Functional differentiation: a critique of the bilingual policy in Singapore. *Journal of Asian Public Policy, 4*(3), 331–341.

Tan, J., & Gopinathan, S. (2000). Education reform in Singapore: towards greater creativity and innovation. *NIRA Review, 7*(3), 5–10.

Tashakkori, A., & Teddlie, C. (2003). *Handbook of Mixed Methods in Social and Behavioral Research*. Thousand Oaks, CA: Sage.

Taskinen, S. (2017). *"Ne voi opita toisilta" – kasvatustieteellinen design-tutkimus maahanmuuttajaoppilaiden osallisuutta edistävistä luokkakäytänteistä* ["They can learn from the others" – An educational design research on classroom practices enhancing participation in immigrant students]. Rovaniemi, Finland: Lapland University Press.

Taskinen, S., Uusiautti, S., & Määttä, K. (2019). How to enhance immigrant students' participation in Arctic schools? In S. Uusiautti & N. Yeasmin (Eds.), *Human Migration in the Arctic – The Past, Present, and Future.* (pp. 143–169). Singapore: Palgrave Macmillan.

Teng, A. (2018, October 24). Nearly half of low-income students in Singapore attend the same schools. *The Straits Times*.

Thalhammer, E., Zucha, V., Enzenhofer, E., Salfinger, B., & Orgis, G. (2001). Attitudes toward minority groups in the European Union: a special analysis of Eurobarometer

2000 survey, Vienna: The European Centre on racism and xenophobia. *Social Psychology*, *41*(4), 230–237.

Thomas, D. A., & Ely, R. (1996). Making differences matter: a new paradigm for managing diversity. *Harvard Business Review*, *74*(5), 79–90.

Uusiautti, S., & Määttä, K. (2013, July 15). Significant trends in the development of Finnish teacher training education programs (1860–2010). *Education Policy Analysis Archives*, *21*(59). Retrieved from http://epaa.asu.edu/ojs/article/view/1276

Uusiautti, S., & Yeasmin, N. (2019). *Human Migration in the Arctic – The Past, Present, and Future*. Singapore: Palgrave Macmillan.

Valtonen, K. (2016). *Social Work and Migration. Immigrant and Refugee Settlement and Integration*. New York: Routledge.

Vasu, N. (2012). Governance through difference in Singapore. *Asian Survey*, *52*(4), 734–753.

Vertovec, S. (2005, September 20). Opinion: super-diversity revealed. *BBC News*. Retrieved from http://news.bbc.co.uk/2/hi/uk_news/4266102.stm

Vertovec, S. (2006). *The Emergence of Super-Diversity in Britain*. Oxford, UK: Centre of Migration, Policy and Society.

Vertovec, S. (2007). Super-diversity and its implications. *Ethnic and Racial Studies*, *29*(6), 1024–1054.

Vuorenkoski, L. (2000). *Childhood between Two Countries. Resilience and Mental Well-Being of Finnish Remigrant Children and Adolescents*. Oulu, Finland: University of Oulu.

Waldow, F., Takayama, K., & Sung, Y.-K. (2014). Rethinking the pattern of external policy referencing: media discourses over the "Asian Tigers". PISA success in Australia, Germany and South Korea. *Comparative Education*, *50*(3), 302–321. doi:10.1080/03 050068.2013.860704

Watson, W. E., Kumar, K., & Michaelson, L. K. (1993). Cultural diversity's impact on interaction process and performance: comparing homogeneous and diverse task groups. *Academy of Management Journal*, *36*, 590–602.

Yeasmin, N. (2018). *The Governance of Immigration Manifests Itself in Those Who Are Being Governed: Economic Integration of Immigrants in Arctic Perspectives*. Rovaniemi, Finland: Lapland University Press.

Yeasmin, N., & Uusiautti, S. (2018). Finland and Singapore, two different top countries of PISA and the challenge of providing equal opportunities to immigrant students. *Journal for Critical Education Policy Studies*, *16*(1), 207–237.

Yingyue, H. (2012, March 16). Singapore as a global city. *The Diplomat*. Retrieved from https://thediplomat.com/2012/03/singapore-as-a-global-city/

Yle. (2017). Positive discrimination funding boosts educational progress among boys, immigrant pupils. Retrieved from https://yle.fi/uutiset/osasto/news/positive_discrimination_funding_boosts_educational_progress_among_boys_immigrant_pupils/9783710

3 Syrian students at the Arctic Circle in Iceland

Kheirie El Hariri, Hermína Gunnþórsdóttir and Markus Meckl

Introduction

In 2016, Iceland welcomed a group of Syrian refugees, who were resettled from Lebanon as quota refugees in collaboration with the United Nations High Commissioner for Refugees (UNHCR). All of the refugees arrived as families with children of school age. A crucial element of the refugees' integration process is education, as it is a key site in which both the host and incoming population learn about one another (Hannah, 2007). As a result, it is important to ameliorate the educational policies and practices targeting refugees. This can be accomplished by studying the educational experiences of the students as well as that of the individuals involved in their educational process, such as the parents and teachers.

This research contributes to the understanding of the experiences of the Syrian refugee children and their respective parents regarding the Icelandic education system, in addition to the experiences of the teachers when dealing with this specific group. This was achieved by assessing the children's, parents' and teachers' experiences at the compulsory school level. So far, only one research paper has been published about the refugee children from Syria in the Icelandic school system (Ragnarsdóttir & Rafik Hama, 2018), which focused on the first year of their arrival. The lack of studies may be related to the fact that Syrian refugees have been in Iceland for only 3 years.

This study was conducted with families residing outside the capital area, close to the Arctic Circle, where there are hardly any other Arab people. The purpose of this research was to assist in improving policies and practices relating to refugees' education. It will help teachers and the school system in general to identify and understand the needs of the refugee parents and students, and will aid the school administration to support collaboration and engage in a mutual dialogue in home–school matters. The findings revealed that the parents lacked trust in the Icelandic education system. They were unable to understand how the all-inclusive system functions, considered that the schools lacked discipline, and viewed the teaching as weak. The parents' perspectives were mainly due to a home–school communication gap, as well as a clash in cultural values regarding the process of education. Both parents and students showed uncertainty about their future in Iceland. Additionally, teachers were underprepared and not supported in their quest to deal with diversity. That said, students revealed feeling comfortable and content regarding the Icelandic schools and had good relationships with their teachers.

This chapter comprises four main sections. The first section presents background information, in addition to a literature review and the theoretical framework. The second section explains the methodology, the methods followed to gather and analyze data, and the ethical considerations. The third section includes the findings of the study, and, in the fourth section, the findings are discussed.

Theories and background information

Icelandic educational system

The Icelandic education system is divided into four stages: pre-primary school education (*leikskóli*), compulsory education (*grunnskóli*), upper secondary education (*framhalsskóli*), and higher education (*háskóli*). The educational policy is based on six fundamental pillars, which are derived from laws governing preschool, compulsory school, and upper secondary school education (Ministry of Education, Science and Culture, 2014). These fundamental pillars are:

1   literacy, which deals with empowering students to create meaning of their own world through reading and writing, as well as technology and media;
2   sustainability, which concentrates on creating active citizens who are interested in local and global issues related to the environment, society, and economy;
3   democracy, which aims to nurture a democratic environment within the school and to create critical citizens who have a vision for the future;
4   equality, which aims to create an inclusive school, to increase students' understandings of various languages, cultures, religions, nationalities, and disabilities, as well as prepare both genders to participate equally in society;
5   health and welfare, which targets creating a safe and positive environment that nurtures the students' welfare and well-being; and
6   creativity, which aims to increase the students' critical thinking, innate curiosity, and entrepreneurial skills (Ministry of Education, Science and Culture, 2014).

The fundamental pillars must be present in all school activities and should be reflected in all subjects and subject areas (Ministry of Education, Science and Culture, 2014). Local municipalities are responsible for the operation and evaluation of compulsory schools (Compulsory School Act No. 91/2008). In 2000, Iceland became part of OECD PISA studies (EURYDICE, n.d.). According to PISA 2015 results (OECD, 2017), students' life satisfaction and sense of belonging to their school in Iceland are above the OECD average, and students' schoolwork-related anxieties are below average. The main findings of a study realized with immigrant students in four Nordic countries revealed that students at Icelandic compulsory schools, aged between 8 and 15, had positive experiences in schools, especially when dealing with their teachers, and specified that they always felt welcomed. These students described their teachers as "caring", "helpful", and "good" (Ragnarsdóttir, 2015).

Literature review and theoretical background

When refugees arrive in a host country, they normally do so in groups, which makes them a minority group with significant characteristics. The extent of their willingness to move to a new culture largely affects the attitudes of acculturation (Anderson, 2003). Berry's acculturation theory is one of the most widely used. He identified four acculturation strategies related to two main issues, which are the extent of one's preference to maintain one's heritage and cultural identity, and the extent of one's preference to engage with the larger community along with other ethno-cultural groups (Berry, 2005). The strategies have different names depending on whether they involve the dominant or non-dominant group. In the case of the refugees (the non-dominant group), assimilation occurs when individuals do not want to maintain their own cultural identity, but rather seek daily interaction with the host culture, whereas separation occurs when individuals hold onto their own heritage and avoid interacting with the other culture (Berry, 2005). In this case, the language, values, and beliefs of the host country are not acquired, but the heritage culture is preserved (Anderson, 2003). Integration occurs when there is an interest in maintaining one's own culture and engaging daily with other groups (Berry, 2005). Integration is considered the most adaptive option accompanied by the most positive outcomes. Lastly, marginalization is the least adaptive option (Anderson, 2003) and happens when there is little interest in either maintaining one's heritage (for reasons of enforced cultural loss) or interacting with other groups (for reasons such as exclusion and discrimination; Berry, 2005). Marginalization is associated with lack of competence for both the heritage and the host language, and with negative attitudes towards both cultures (Anderson, 2003).

One of the main acculturating agents within societies are schools, as they transmit the values, norms, and tools of a certain culture; this can include multiculturalism in addition to attitudes and beliefs about specific migrant groups. Additionally, they are the prime contact between the immigrant and the host community. Two factors that can affect the process of adaptation and adjustment in schools are the expectations of how to behave in schools and how similar or different the cultural frames are (Anderson, 2003). In the case of the refugees, they are more likely to find themselves in countries with little cultural similarity to their own, as they have no choice in choosing the country of resettlement (Sheikh & Anderson, 2018).

Multiculturalism is here understood as an idea that values equality, encourages cultural identity to thrive, and ensures that no specific group dominates the others (Castles, 2009). Schools can be viewed as a microculture constituting of a dominant culture and subcultures and having a set of norms, values, and goals (Banks, 2009); the process of acculturation also occurs within schools, and the values of multiculturalism can be applied in schools under the title of "multicultural education".

Multicultural education is a response to the limitations of assimilation and segregation forms of schooling for students from various migrant backgrounds. It seeks to fully develop their potentials and acknowledges their differences by combining both the principles of recognizing cultural differences and working towards equality (Castles, 2009). Multicultural education emerged with the rise of the 1960s civil rights movement, when African Americans demanded their rights in the United States. One of this movement's major goals was to eliminate discrimination in education systems by employing more black and brown teachers and administrators, revising textbooks to reflect diversity, and reforming the curricula to reflect their experiences (Banks, 2010).

James Banks is a recognized scholar in the field of multicultural education. He defines multicultural education as "a total school reform effort designed to increase educational equity for a range of cultural, ethnic, and economic groups" (Banks, 2010: 7). The Icelandic education policy follows the concept of inclusion (Ministry of Education, Science and Culture, 2014). According to international comparison, Iceland is considered highly inclusive in its education system, with very limited segregated resources for students with special needs (Gunnþórsdóttir & Jóhannesson, 2014). Article 2 of the Icelandic Compulsory Act places emphasis on the general development of all pupils, and Article 17 mentions that pupils with special needs are permitted to have their educational needs met in a regular inclusive school (Compulsory School Act No 91/2008). Inclusion is an integral part of multicultural education as it aims to celebrate diversity and ensure the participation and success of all students who face any kind of learning and/or behavioral challenges with regard to socio-economic situations, cultural background, ethnic origin, sexual preference, religion, gender, and so on (Topping & Maloney, 2005). Initially, inclusive education was about incorporating students with disabilities into mainstream classrooms, yet, recently, it has also been responding to the increasing diversity, such as cultural and linguistic, within schools' communities (Taylor & Sidhu, 2012). The Icelandic compulsory national curriculum defines inclusive schools as those who consider all students as having equal opportunities for education and who work towards meeting these students' specific educational and social needs. Following this definition of "inclusive education", the equity pedagogy dimension of multicultural education plays an important role. In Iceland, equity is achieved by individual-oriented teaching methods and by taking into account the values of equality and the needs and experiences of individual pupils (Ministry of Education, Science and Culture, 2014).

One way to enhance cultural understanding is to follow Hofstede's work on national culture dimensions (Hofstede, 2011). Hofstede's national culture dimensions describe the effects of a society's culture on the values of its members, and how these values appear in behavior. Hofstede's study is a useful tool to understand and compare the participants' responses regarding the education system and the learning environment in relation to culture. Hofstede identified six dimensions in order to understand the cultural values of people belonging to different countries (Hofstede, 2011); they are: power distance, indulgence/restraint, masculinity/femininity, collectivism/individualism, long/short-term orientation, and uncertainty avoidance.

Good examples to understand how Hofstede's national culture dimensions can be implemented in education are the power distance and masculinity/femininity dimensions. In countries with large power distance, education is teacher-centered, the relationship between teachers and students is formal, and teachers are treated with respect or fear (Hofstede et al., 2010), whereas, in countries with small power distance, education is student-centered, teachers and students are considered equals, and learning is a two-way communication process (Hofstede et al., 2010). In masculine societies, there is high competition between students, high achievers are rewarded, winning is important, and failure is not accepted. In contrast, in feminine societies, competition is not widely approved, failure is fine, and there is praise for the weak (Hofstede, 2011; Hofstede et al., 2010).

The development of clear communication channels between home and school can ease the clash of cultural values and expectations when refugees start school. Establishing clear communication channels is directly proportional to increasing parents' involvement in schools (Hamilton, 2003). Various literature and studies emphasize the important role that home–school collaboration and parental involvement play in the success of refugee children (Block, Cross, Riggs, & Gibbs, 2014; Ficarra, 2017; McBrien, 2005; Taylor & Sidhu, 2012; Thomas, 2016). Additionally, one of the significant factors in achieving school reform and multicultural education is the involvement of the closest people to the students in the teaching and learning process (Banks, 2010; Nieto & Bode, 2010).

Methodology

Participants

Three Syrian families, comprising three fathers, three mothers, and six children, were recruited voluntarily based on the following criteria: (1) identify as a refugee per quota, (2) reside outside the capital area, and (3) have at least one child registered in elementary school in Iceland. All of the fathers had reached upper-level education in their home country, whereas the mothers had limited education – either informal or elementary level. All of the fathers and two of the mothers were employed. The children, one girl and five boys, were between 8 and 15 years old. The parents were first approached informally, either face to face or via a phone call, where the research objectives were discussed, and their oral agreement to take part in the study was gained. After their oral agreement, the parents were formally presented with an Arabic version of a letter of intent and an informed consent document. The children were only interviewed after giving their own personal consent and their parents' having consented. Interviews were conducted in the families' houses in a room separate from other family members. Parents were interviewed separately, whereas siblings were interviewed together.

School principals were asked to identify teachers who were responsible for Syrian students at the compulsory level. Five teachers, four females and one male, were interviewed. Only one teacher had previous experience with refugee teaching in a non-formal context – that is, teaching students who are not registered in formal schools. Prior to contacting school principals, consent was obtained from the director of school authorities. The teachers were provided with an English version of the letter of intent and informed consent. The interview was conducted in a location chosen according to the teachers' preference – either in a school classroom or in a café.

Data gathering and analysis

In this research, semi-structured interviews were conducted, as they serve the objective of examining the experiences of the participants by allowing them to express their opinions and ideas in their own words (Esterberg, 2002). Semi-structured interviews, as opposed to structured interviews, allow for open-ended questions, which gives the interviewees the chance to shape their responses or change the direction of the interview altogether (Fife, 2005).

All participants were interviewed once only in spring 2018. One-on-one interviews were conducted with the teachers and parents, whereas group interviews were conducted with the students, as children tend to get shy and overwhelmed in one-on-one interviews (Fife, 2005). Siblings of the same family were grouped together, and groups ranged from two to three individuals. A one-on-one interview was conducted with one student as he was the only child in his family to fit the required age range to be a participant in this study. Parents' and students' interviews were conducted in Arabic, and the teachers' interviews were conducted in English.

Interviews were first conducted with the families. Parents were mainly asked general, open-ended questions, such as: "What would you like to improve in the school?" and "Describe the ideal school for your child". Children were asked similar questions, but in a simpler form, and drawings and scales were used in order to facilitate their answering process. For example, students were asked to choose from a variety of faces (e.g., happy face or sad face) as an indicator of their level of happiness in school and they were also asked to draw what their ideal classroom would look like (e.g., seat arrangement).

The teachers' interview questions were set according to the results of the parents' and the students' answers. Therefore, the questions were more specific, such as: "What do you think about the concept of homework?" and "How do you motivate the students?" They were also asked general questions, some of which were: "What were your expectations before teaching the student(s) and after teaching them?" and "What are the challenges, if any, that you encountered while teaching the student(s)?"

All interviews lasted between 30 and 60 minutes, and they were audio-recorded with the agreement of the participants. Following each interview, the audio-recording was copied and transcribed in English. The parents' and students' interviews, which were in Arabic, were directly translated and transcribed in English. Some limitations occurred when trying to preserve certain meanings – for example, Arabic proverbs. This limitation was handled by writing notes to clarify the meaning. Utterances (e.g., "um", "uh") and repeated words were frequently removed, and aspects such as jokes, laughter, long pauses, and external interruptions were noted.

The data were analyzed using thematic analysis as it permits the reflection of participants' realities (Braun & Clarke, 2006) and treats data in a descriptive manner (Vaismoradi, Turunen, & Bondas, 2013), thus serving the objective of this research. Thematic analysis systematically identifies and describes themes (patterns) across data. Themes represent a specific pattern found in the data that serves the research question(s) (Braun & Clarke, 2006; Joffe & Yardley, 2004).

Ethical considerations

The interviews started once all ambiguities had been clarified and signatures had been obtained. The children were not interviewed unless their parents consented. Both the parents' and the children's signatures (or in this case a written name) were needed to start the interview. The children were also informed about the research objectives and interview procedures beforehand.

School principals were contacted once approval from the director of school authorities was obtained. Similarly, teachers were contacted with the agreement of their school principals, who provided me with their contact information. As in the case of the parents and the children, interviews with the teachers were conducted after all their questions regarding the research had been truthfully answered, and after they had signed the informed consent.

All participants were informed that their participation was voluntary, and that they were allowed to terminate it at any time. They were also notified about the audio-recording. A hard copy of the informed consent and letter of intent was left with the parents, and we retained a soft copy. In the teachers' case, it was the opposite: the informed consent was scanned and sent to them via email. Prior to the interviews, the teachers were also provided, via email, with the letter of intent.

In the findings, we were very careful not to include any data that would identify the participants: pseudonyms were used. Also, we avoided mentioning details such as school names, period lived in Lebanon, and the city of birth. For the sake of anonymity, we considered all students as males and all teachers as females, as gender was not a variable in this study. Parents were referred to using pseudo family names: Mr. and Mrs. Zain, Mr. and Mrs. Loutfi, and Mr. and Mrs. Faraj. Both students and teachers were given pseudo names. The names given to the students were: Nadim, Wael, Adam, Bilal, Alaa, and Moussa; the names given to the teachers were: Dina, Lisa, Kristin, Emma, and Dalia.

Findings

This chapter presents the research findings, which are divided into four main themes: (1) the Icelandic school system, (2) the parent–school relationship, (3) supporting teachers and staff, and (4) is Iceland in the future? The first theme is made up of four subthemes: "individualized teaching and students' academic achievements", "subject area: the value of arts and crafts", "discipline", and "homework".

The Icelandic school system

Individualized teaching and students' academic achievements

In Iceland, the education system is based on inclusivity, and the teaching is individual-based. Students have their unique set of objectives to achieve throughout the year and are given specific material adapted to their level. Throughout the interviews with the parents, it was obvious that they did not understand how the all-inclusive system works. The choice of having all students together in one classroom, despite their differences, was not understood, for example, by Mrs. Loutfi:

> Now they put them [the Syrian students] only 15 days together, then they took these students to Icelanders class. He [her son] entered to class very lost. How will he speak?

The parents, being used to a traditional evaluation system, were very confused regarding a system where individual objectives are set for each student. Mr. Loutfi demanded that he "needs to see results of his children" and wondered that if "there is something called quiz, test and exam [...] where are the results?" Mr. Faraj doubted the teachers' evaluation by saying:

> We always ask about the children; they [the teachers] tell us all is very very good. But the truth is they [his children] are zero [have very low academic level], so only if they [the teachers] do an evaluation, not only to encourage us, a true evaluation to specify each student's level if that is possible.

As no numeric grading system is used, the parents feel lost in supervising the progress of their children. Furthermore, contrary to the idea of inclusiveness, parents are stressing the importance of competition to motivate students to work harder. Mrs. Faraj explained this by saying:

> We [referring to the Syrian and Arab society] have the sense of competition, for example, if one student takes a grade of 10 then the other student will want to receive the same grade [...] so, the child will want to work hard and learn in home.

Similarly, Mr. Zain mentioned this point, comparing it with his home country:

> we had this in Syria, for example, reward the first ten then those who are rewarded will always improve and their friends will try and keep up with the top students [...] And this they do not have it here.

The parents perceive the lack of competition as a weakness of the system, as students cannot fail a school year or, as Mr. Loutfi expressed, "there is no evaluation" because "the student whatever he is, he is passing [moving to a higher grade]".

The children, however, did not mention any disturbances or challenges concerning this aspect. They considered their teachers helpful. When asked about how their relationship was with their teachers, some of the responses were:

Bilal: Good, if I want help they help me.
Moussa: Their treatment is very good [...] they come alone to me and help me, if I need anything.

Likewise, when Alaa was presented with a scale of smiley faces and asked to point to the face that indicated his level of contentment with his teachers, he chose the very happy face. His explanation for this was "[because] they help me", or, as Nadim said, "they treat me in a very good way". Additionally, some of the parents appreciated this explicitly, as Mr. Faraj pointed out:

> For example, they take into consideration if he [the Syrian student] is sick, or has a psychological problem [...] they [the teachers] do not leave the student alone by himself [...] They find for him a solution.

Even though the parents appreciated that the school system does not discriminate against their children, the insufficient understanding concerning the function of the all-inclusive system created dissatisfaction among the parents. They worried that their children were not getting the academic support they needed. This confusion indicates that there is a lack of effective home–school communication and a failure to clearly explain how the system works to the parents.

The main reasons for the parents' confusion seem to be the language barrier and the teachers' lack of preparation. The teachers indicated that the goals are in Icelandic, in which the parents have limited proficiency. Furthermore, as Dalia mentioned, the teachers are still "adapting" to the system of setting individual goals. Kristin mentioned that, "the new system [of setting objectives] [...] is very hard to work on", and that the teachers are confused about how to execute the policy. These two aspects prevent successful communication.

Subject area: the value of arts and crafts

In the compulsory curriculum in Iceland, the weekly proportion of arts and crafts is 15.48 percent, which is more than mathematics (14.88 percent) and natural sciences (8.33 percent). Arts and crafts are defined as follows: "to arts belong music and visual arts and dramatic art. To crafts belong design and handicraft, textiles and home economics" (Ministry of Education, Science and Culture, 2014: 51).

While discussing the subject area with the parents, it was obvious that there is a lack of efficient communication between home and school. Although there is emphasis on arts and crafts in the curriculum, parents were not able to ascertain the importance of these subjects. Mrs. Loutfi indicated that, "these subjects [arts and crafts] we do not care about", considered that there are "a lot of missing subjects", and wished that her children would take more "reading and writing".

Furthermore, Mr. Loutfi assumed that these subjects are "activities" that "will not benefit anything" and aimed for subjects that would provide more "knowledge", such as learning the English language. Similarly, Mr. Zain stated that, "the scientific subjects it seems they [schools in Iceland] do not have that interest in them". He defended his statement by comparing the school with that in Syria, by saying:

> I am talking about the level in Syria – physics and chemistry these were nice and they used to concentrate on them a lot. Now, the level of my children is less […] they are still young. But [for example] in geography and geographic coordinates […] put the map and tell the child where is Britain he will not know whereas in Syria we used from the class – age of my son they used to put the map and tell us this is Egypt […] We knew how to locate them on the map. Now I am sure no one knows how to locate anything on the map.

Teachers, on the contrary, highly valued subjects belonging to the category of arts and crafts (e.g., carpentry) and stressed their importance to increase creativity and teach self-care. Kristin explained by saying:

> We think it is very important to teach them that [arts and crafts]. By that we teach them to be creative, to make something, to get some ideas, not to be just like computers what the teachers tell you. Be creative. Think on your own.

Additionally, teachers emphasized the value of arts and crafts in order to discover one's self. Lisa elaborated on that, by saying:

> students are just people who have really different strengths. I have got students that are just unable to do math, they do not have the capability to do it, it is really hard, but at the same time they are excellent at sports or dancing or something. It is just that people have different strengths, and in elementary school it is important to both get to work with your mind and your hands. To just find out where your strengths are.

The teachers' opinions reflect what is stressed in the compulsory national curriculum guide. Creativity is stated as one of the six pillars upon which the working methods, communication, and schools' atmosphere are based. Additionally, the curriculum highlights the importance of creating independent individuals and developing the talents and abilities of each student (Ministry of Education, Science and Culture, 2014). The parents failed to recognize these characteristics and to appreciate the value of these subjects for their children.

All students enjoyed the subjects of arts and crafts. None of them made any negative comments about these subjects. For example, Nadim expressed that his ideal school is "like our school [current school]". When asked why, he explained:

> There is sewing, carpentry, drawing and like that. Things like clay. There is music, sports. Whatever you want there is. There is cooking.

Similarly, Alaa pointed out that what motivates him to study is "carpentry and sewing", and Wael expressed how he liked his current school by comparing it with schools in Syria: "Not like Syria. Syria [is] all studying. There is no sewing, carpentry".

Discipline

Discipline is another aspect that shows that the parents have a very different perception of what a school is supposed to be compared with those found in Iceland. The Syrian parents considered that the Icelandic schools lacked discipline and order, and that the students showed little respect towards their teachers. For instance, Mrs. Faraj mentioned that they "do not feel the order" and compared the school with those in Syria and in Lebanon, by saying:

> it is not like in Lebanon [...] there is order [in Lebanon], the child is clean and neat. He goes to school wearing the school's uniform and his hair neat and combed, his nails cut and clothes ironed and clean. Here there is nothing of this order.

Additionally, her husband, Mr. Faraj, expressed that the school system in Iceland is "very much free" in comparison with Syria, where there is "strictness". When asked what he meant by strictness, he explained:

> For example, if the student is not prepared, he knows that he will get punished. So, he works more and he will care more. Here, for example, prepared or not prepared he knows that the result is the same.

A recurrent example that several parents gave to describe the supposed disrespect of the students towards the teachers was similar to this quote from Mrs. Faraj:

> In our country the student sits respectfully and with manners, here the student sits and puts his legs on the table in front of the teacher, and in front of his friend. This is normal to them. We do not have this. In our country, the student [...] have to sit with manners and concentrate on the teacher while she or he is explaining the lesson.

Only one student, Moussa, discussed the perceived lack of discipline in the interviews. Moussa expressed that "the teaching" is a point of improvement in the school. When asked to elaborate on that, he responded by, "for example, while we are studying there is one student playing, another on the phone, and another not writing". He also stated that he missed "the hitting [corporal punishment]" in Syria, because this way "the person will learn directly".

Contrary to this student's response, all the other students indicated that they prefer the teachers in Iceland over those in Syria and in Lebanon, where there was "hitting". Nadim expressed that he "used to hate something called school", because it was "all studying and there was hitting". Also, Bilal explained that he really liked his current teacher because "she does not get angry and scream".

Teachers were aware of the parents' desire to have a "traditional" way of teaching. However, they strongly opposed it, and strove towards building trust and creating open communication with the students. Kristin expressed her opinion regarding this point by saying:

> we are trying to [...] make them feel good. Make them trust us, and make them feel welcomed [...] for the kids of course if it does not come from themselves, they will not learn it, it is very hard to make some [students] to learn something they do not want to. [...] My opinion is that try to make a good nice person, do not make people obey by being afraid of you. Make them want to follow the rules, because it is better.

Similarly, Lisa expressed her opposition to the parents' opinion regarding discipline and emphasized communication to solve conflict between students by stating:

> They [the Syrian parents] think we are not strict enough when it comes to discipline, and they think that we did not address it good enough when he [the Syrian student] has been bullied or something. But, I would disagree, because my student he had the attitude if someone was messing with him, or someone was hitting, he was allowed to hit back, but we are trying to imprint on him that we try to use words and we try to talk about it to solve conflict, not by hitting back. And I think to some degree the parents were okay that he just hit back, and that was a normal way to solve a conflict.

The differing points of view between the parents and the teachers reflect the differing opinions regarding the education process. As in the subtheme "individualized teaching and students' academic achievements", the parents lean towards using external motivation techniques to push the students to work more, such as punishments (Ryan & Deci, 2000). This was also seen in the responses of one of the students, Moussa, whereas the teachers lean towards intrinsic motivation, such as working because of personal interest and joy (Ryan & Deci, 2000).

Homework

All of the parents expressed their desire for their children to have homework. They believe homework is necessary to stay up to date with the academic life of their children and to be assured that their children are learning. For example, Mrs. Faraj mentioned that homework is important to make sure that "the idea will be planted in his [the student's] mind". Mrs. Loutfi stated that it is vital in order to "follow up" and compared the situation to that in her home country and first asylum country, saying, "they come home [with] no books [...] They do not come and write, like in Lebanon and Syria. [There] you see all children are writing".

In addition, she stated that several times she asked the school for homework, but, according to her, "they are not giving them". She then added, "they said that here in Iceland that's it. It [the study] is enough in school, and at home they [the students] do other activities."

The teachers emphasized the importance of students spending their after-school hours with their families and in recreation. Lisa explained this by comparing school to a regular job. She said:

> we have [the] opinion [that] this is their [the students'] work, and because I do not want to take my work home, I do not send them home with their work.

Although parents demanded that their children receive homework, teachers' interviews revealed that teachers do in fact allocate some tasks for the students to do at home. Teachers indicated that students are expected to finish work that is not completed in the classroom. Furthermore, all teachers pointed out that the children have to read at home, where Dalia mentioned that they "expect the parents to let them [the students] read aloud every day", and Kristin stated that, "the parents have to sign their names [to say that] they listened to them read". However, it seems that, according to the parents, this task is not considered homework.

Parent–school relationship

The parents' and the teachers' interviews revealed that communication between the parents and the school is very limited. Parents only interacted with the teachers and/or the administrative staff upon request or during the regular parent–teacher meetings, which occur twice a year. Two families stated that they remained in contact with the school through emails, whereas one family, the Faraj family, mentioned that they had not been receiving emails. In this situation, Mrs. Faraj indicated that they stay up to date through the "emails of their children"; for example, "If there is a vacation, they send [the information] to the children's emails".

Although teachers updated the parents through emails, the parents did not show any responsiveness to these updates. Teachers mentioned that the parents do not reply to the emails and they doubted that they read them. For instance, Dina stated: "they [the Syrian parents] do not read post [emails] from me, even though I have them in English".

When asked if she tried to contact a translator, she replied that she "tried twice"; however, she considered that the father "actually does not know how to read the email, even if he have the computer and number [access] of the email [...] but he doesn't do it". Also, Lisa mentioned that she "always writes them [the Syrian parents] in English" and, thus, assumed that they are reading the emails even though she does not get any responses.

Emma doubted that the parents understood what is written in the emails, saying:

> you send your usual email, but you do not know if they understand. You have these two meetings a year. We get the translator. But, some people are more motivated to try to just Google Translate [to] do something [...] and some people just [do] nothing.

Therefore, the teachers seem to think that the lack of communication lies only with the parents. Emma showed her desire for the parents to be more involved with the school and related the lack of involvement to the lack of language proficiency. This was reflected in the following exchange:

Emma:	we think maybe they should come more to the school. Maybe they do not understand what we are writing them. I do not know.
Interviewer:	Do you send them in the email to come to the schools?
Emma:	Yes, if there is presentation or something. Parent meeting. They are not sure exactly what to do.

Kristin considered the differences in cultural values as a barrier to communication. She highlighted this issue when discussing techniques to enhance parents' understanding of the Icelandic system, by claiming:

It is no use for me to try to make him [Syrian father] understand […] Our translator he is very good in making him understand our rules, and it is much better for him to do it, because he is a man. That is also difference of the culture […] the man is the boss at home. It is not how we do it in Iceland, we try to be all equals, but when a man says something to a man like that, it is better for the man to tell him. He is not going to listen to me I know it. So I just smile.

As a result, this shows that the teachers believe that the barrier to communication is one-sided and related to the parents. In this case, it is due to the culture and lack of language proficiency. No valid solutions were presented by the teachers to overcome these barriers.

Mothers were more passive than the fathers in their interactions with the teachers. All of the mothers mentioned that the fathers are responsible for the emails. This might be related to their limited literacy and lack of proficiency in English and Icelandic. In addition, as seen in the exchange below, teachers noticed that the mothers rarely interacted in the meetings, even when a translator was present.

Interviewer:	I hear that you mostly say father, you do not talk about the mother. Is the mother not involved?
Dina:	She always came also, but she doesn't say much.
Interviewer:	Do you think it is because of the language?
Dina:	Maybe, because she doesn't speak English, and the father speaks English. But the translator always speaks Arabic. So I am not quite sure.

In addition, no extra efforts, other than assigning a translator, were made by the teachers or the schools as a whole to enhance communication with the parents. When asked if the school took any additional steps to get the parents involved, Emma replied, "no". She then indicated that there was only one person, called

Anna, who "took care of them [the Syrian families]", and that the refugees' issues were "just her responsibility". She explained that Anna is responsible for all refugees and is assigned by the municipality. She considered Anna a "communicator", to whom they send an email when they face difficulties with the refugees and who helps them in allocating academic material. Kristin mentioned that parents are always welcome to the school, by stating:

> The parents can see what we are doing in schools, they are always welcome […] My door is always open, and we invite the parents to come to see what we are doing […] and they do not always come. In the beginning they always came. But, sometimes they do not.

Nevertheless, she did not indicate any specific technique to motivate the parents to get involved, and the level of parental involvement is dependent on the parents' willingness. Finally, all of the parents stated that they were not in contact with any of their children's classmates' parents, unless they were one of the Syrian families. Only one teacher, Emma, mentioned the importance of being in contact with other parents, by saying:

> We think maybe because they are trying to get into the community that they should maybe come. And, as well as friends after school for children, you maybe have to get to know the other parents. And, that's why you have to come to school.

Similarly, as in the previous cases of this theme, the responsibility of the parents to be involved is put solely upon them. The teacher expressed her desire that the parents would visit the school more often and communicate with other Icelandic parents; however, she does not indicate any specific actions on how to involve them.

Supporting teachers and staff

The interviews with the teachers indicated that they needed extra support when dealing with the Syrian refugee students. Their aspiration for support varied between providing: (1) psychological help for students, (2) cross-cultural education for teachers, and (3) in-classroom assistance.

Two teachers mentioned the issue of providing psychological help for the refugee students. Lisa expressed her surprise that psychological assistance is not obligatory for refugees, saying:

> we have put pressure on […] the Red Cross and the local government that this kid [the Syrian student] needed more help than we as a school can give. And, I actually think it is really strange that children that come here from conflict area, it is not an obligation that they should get psychological help.

Dalia criticized the school regarding this aspect, stating:

> This is a school of about 500 kids, and I think they should have a children psychologist, and they have to have someone who is specialized in therapy [...] if there are moving problems. Physical and psychological. And, provide more service. More special teachers.

Teachers also aspired to learn more about the culture. When asked what mostly attracted her about the Syrian culture, Emma replied: "I do not know. Maybe we do not know too much about the culture. Maybe we should do more learning". In addition to her wish to learn the culture, she aspired to "learn a little bit about Arabic, and the language".

Additionally, teachers discussed the pressure they felt regarding all-inclusive classrooms and desired to have extra help in preparing academic material and in teaching the students. Lisa discussed the negative point of all-inclusive classrooms, stating:

> the bad thing is that now I got in this class alone four children that need a lot of extra help, and I am not getting extra help too to work with them.

She later gave an example of a different school, in Norway:

> there [in Norway] were always two teachers in the classroom, because there are always some students who need extra help. And because in Icelandic I am working with four different types of material, it is hard to be teaching something up here when they are doing something completely different.

Only one student raised the issue of teacher support within the classroom. He mentioned that the "studying is weak", because there "are 28 individuals [students], and she is only one teacher". He later gave a solution for that by stating: "that is why she needs another teacher to help her, so that everyone understands".

Is Iceland in the future?

When parents were asked if they envision their children going to Icelandic universities, they either showed uncertainty or answered by a direct "no". The main reason for this is related to their lack of trust in the Icelandic education system. For example, Mrs. Loutfi emphasized that her children will not be going to Icelandic universities by saying: "I [...] talked with a journalist [...] about this. He asked me if my children will base themselves here in universities [...], I told him no". Her husband, Mr. Loutfi, voiced his worries about the uncertainty of the future of his children, related to the lack of planning found in the Icelandic schools, arguing:

> When will they [his children] reach university, we do not know, because the holder of the plan [meaning the teacher or principal] does not know [...] Now [if] I go [...] to the upper secondary school, and ask him [the teacher]

when will they finish, with all honesty he tells you he does not know [...] It is supposed that when you bring a refugee [...] it is known [that in] three years he [the student] will finish this level, four years this level then five years this level, so that he [the student] can know how to plan his life.

Additionally, when asked where he sees his children 5–7 years from now, Mr. Faraj claimed that "in Iceland [it is] difficult", and explained the reason by stating:

for the schools, for the language, for everything. Here they say they are fifth worldwide for teaching [quality], but we do not see this. For example, the doctors [in Iceland] they have all learned in Canada, in America, in Britain.

The parents' desire to have their children continue their higher education in countries other than Iceland explains their emphasis on the importance of learning the English language. Mrs. Zain mentioned that, "they [her children] should take English" because "this language is the best wherever you go". Furthermore, Mrs. Loutfi voiced her complaints regarding the lack of focus on the English language in the school, stating: "now my son he is talking in school English; they do not want him to speak English in the school. They want Icelandic". She later expressed that English is important because it is "the language of the whole world".

The teachers were conscious of the parents' desire to eventually leave Iceland. Nevertheless, they did not identify the lack of trust in the Icelandic schooling system as the main reason for their desire to leave, but, rather, the effect of the language and the surrounding community. For example, Lisa pointed out that the reason why the Syrian families want to leave is because "the Arab community is not [...] big", and they aspire to be in a country where "they have [...] relatives". In addition, Emma indicated that the Syrian families might want to go to a country where "they can speak English", such as England.

All of the students except one expressed their wish to move to another country. The students did not widely elaborate on the reasons why they wanted to move to another country. It is well known that the beliefs of the parents affect the children (McBrien, 2005), which might lead the children to have similar aspirations to those of their parents. Only two of the students presented reasons for their desire to leave, which were related to the language and the environment. Alaa voiced his desire to move to "Australia", as there "they [the people] speak English", and Nadim explained that he does not want to remain in Iceland because it is "cold".

Discussion

Three major themes emerged from the findings in the light of the theories and literature presented in this research. These themes are: the cultural differences, the communication gap, and educational support.

The effect of cultural values on education: Iceland versus Syria

It was apparent that the parents have expectations of the education system that are in opposition to the practices found within the schools in Iceland. These expectations sprang from their home country's culture, which the parents usually compared with the education system in Iceland. Students also compared the schools to those in their home country; however, in contrast to their parents, the students preferred the Icelandic schools. They favored the treatment by the Icelandic teachers, whom they considered less strict and more caring, and they enjoyed subjects such as arts and crafts, whereas the parents failed to understand and appreciate the importance of such subjects.

Parents desired a learning environment where there is discipline and order. They perceived the Icelandic schools as very "free" and considered the teachers not strict enough. In addition, they wished for a system that encourages competition between students and supported the idea of failure – that is, class competition. Parents wanted the high achievers to be rewarded and/or have a certain punishment system as a motivation to learn. This was also reflected in the aspirations of one of the students, who demanded the presence of corporal punishment to push students to learn more. In contrast, the teachers stressed the importance of internal motivation, such as working towards a future goal or studying because of intrinsic enjoyment. They opposed the ideas of performance and competition, empathized with the situation of each student, and considered it unfair to compare students with one another. Additionally, teachers stressed the importance of open communication to resolve problems, rather than merely punishing, and strongly rejected the idea of corporal punishment.

These oppositional points of view regarding the learning environment can be explained through Hofstede's cultural dimensions. Iceland has low masculinity and small power distance in comparison with Syria, which has high masculinity and large power distance (Hofstede Insights, n.d.). According to Hofstede et al. (2010), in feminine societies (i.e., countries with low masculinity), failure is accepted, competition is not openly encouraged, and there is praise for the weak, whereas, in masculine societies, competition is encouraged, failure is not accepted, and the concept of rewarding high achievers is endorsed (Hofstede et al., 2010). Furthermore, in countries with small power distance, the learning process is a two-way communication, students and teachers are considered equals, and corporal punishment is considered child abuse. On the other hand, in countries with large power distance, teachers are usually treated with respect and fear, corporal punishment is accepted, and strict order is expected to be found in classrooms (Hofstede et al., 2010).

Besides the learning environment, parents showed dissatisfaction regarding subject areas in Iceland. They considered that, "a lot of subjects [were] missing", and that the Icelandic academic level was poor. They also believed that subjects falling under the category of arts and crafts were just "extra-curricular activities" and wished that the schools would concentrate more on subjects that provide "knowledge". Teachers, on the other hand, highly valued these subjects and emphasized their importance in increasing students' creativity and independence, while assisting students in discovering their strengths. This is also emphasized in the compulsory national curriculum, where creativity is considered one of the six pillars upon which the education policy is based. Schools are expected to increase the students' independent thinking and self-responsibility, in addition to stimulating their imagination and curiosity (Ministry of Education, Science and Culture, 2014). As students of a low power distance country, Iceland, students are expected to be independent and discover their own intellectual paths (Hofstede et al., 2010).

Syria is a country with strong uncertainty avoidance, and, thus, individuals belonging to such societies need structure, clarity, and rules and are not comfortable with unpredictable and ambiguous situations (Hofstede, 2011; Hofstede et al., 2010). This may explain the demands of the parents for homework, where their main reason for that is to "follow up" and make sure that their children are studying. It likewise explains the anxiety of the father, found in the section "Is Iceland in the future?", concerning the uncertainty of his children's future, and his frustration with the lack of clear, long-term planning from the Icelandic teachers and schools.

The findings discussed in this section reflect how a society's culture can affect the process of education and individuals' beliefs and values. However, it is important to keep in mind that there is no "good" or "bad" culture, as culture symbolizes values that are the creation of historical and social conditions and necessities. Culture is dynamic, meaning that it is constantly evolving as a result of political, social, and other factors in the immediate environment (Nieto, 2009). Therefore, Hofstede's cultural dimensions are used to enhance cultural understanding (Hofstede, 2011), and one should be critical while avoiding creating stereotypes when referring to them.

Having oppositional cultural frames can be considered a reason for the parents' lack of trust in the Icelandic system. It may also lead to a negative acculturation process, where the refugee families can become either separated or marginalized (Berry, 2005; Sheikh & Anderson, 2018). This is reflected in some of the parents' lack of interest in the Icelandic language, especially given that they do not see themselves living in Iceland in the long term. Additionally, refugee parental beliefs can affect the children (McBrien, 2005), which may explain the children's aspirations to move to other countries.

Schools, being one of the vital agents in facilitating the acculturation process, need to establish appropriate actions to foster integration and decrease the negative influence of poor cultural fit. According to Berry, following the multicultural approach is the most appropriate way to achieve successful integration (Berry, 2005). Teachers need to be trained in how to deal with diversity (Anderson, 2003) and learn about the refugees' cultural background. However, the results of this study indicate that the teachers have insufficient knowledge about the refugees' culture, which may impede the successful educational resettlement of the refugee children (Lerner, 2012). Furthermore, establishing clear communication channels between home and school is an essential part of multicultural education (McGee Banks, 2010) and can ease the cultural clash (Hamilton, 2003).

A home–school communication gap

One way to overcome differences of prior conceptions, values, and goals between the parents and schools is to establish clear home–school communication (Hamilton, 2003). The findings in this study show that there is a wide home–school communication gap, as described in "Parent–school relationship", as well as the subthemes "Homework" and "Individualized teaching and students' academic achievements".

Hamilton discussed the importance of parental involvement in building clear communication channels between home and school. He described the relationship between parental involvement and home–school communication as reciprocal, where increased parental involvement develops a solid communication channel, and, in return, a clear communication channel enhances parental involvement (Hamilton, 2003). Richman pointed out that the more parents are involved in the school, the more they will be knowledgeable about and comfortable with the school (as cited in Hamilton, 2003). The findings presented in the "Parent–school relationship" theme reveal that the parents' involvement in the school is limited, which may be one of the reasons parents lack trust in the Icelandic school system.

Several studies and literature reviews indicate that one of the main causes for limited parental involvement is the language barrier. This is reflected in the findings, where the teachers doubted that the parents understand or "know how to read" the emails sent. Thus, schools are advised to establish language training programs for the parents. In addition to language, computer skill training programs can enhance the level of parental involvement (Hamilton, 2003; McBrien, 2011; Rah, Choi, & SuongThi Nguyen, 2009). The results indicate that the parents have limited understanding of the email and computer system, where information about the school and students is regularly communicated. One of the families pointed out that they do not have access to their email account, and that they depend on their child's email account to get information about the school. Moreover, sending emails is considered a one-way form of communication, and, if the parents have difficulties with the language, they will miss opportunities to inquire about the school and their children's performance (Gunnþórsdóttir, Barillé, & Meckl, 2018).

Part of parental involvement includes volunteering in children's schools and classrooms, as well as being part of parents' associations and councils (Rah et al., 2009). These activities might serve as a solution to one of the teachers' aspirations, which is for the parents to have greater contact with other parents, especially given that schools are considered an essential contact between immigrants and the host culture (Anderson, 2003). However, all of the Syrian parents specified that they are only in contact with each other, putting them at risk of separation or marginalization (Berry, 2005).

The findings also show that the mothers' level of involvement is more limited than the fathers'. Teachers pointed out that the mothers rarely shared their opinions in the meetings, even in the presence of a translator. Additionally, in all the families, the fathers are the ones responsible for checking their emails.

This might be related to the mothers' limited literacy, or to the culture in masculine societies, such as Syria, where women expect male dominance. Women are characterized by their gentleness and care, whereas, men are characterized by their sense of responsibility, decisiveness, and ambitiousness (Hofstede et al., 2010). Hamilton stated that, "depending on the target population, the definition of 'involvement' may differ dramatically such that it is important that schools do not adopt the same expectations for involvement for all parents within the school, irrespective of their needs" (Hamilton, 2003: 93). As a result, schools need to be culturally sensitive in their efforts to involve parents; however, this does not seem to be the case here. One of the teachers openly discussed her inability to communicate with Syrian parents due to a clash in cultural values. She assumed that one of the fathers would not listen to her because she is a woman and, consequently, did not increase her efforts to enhance communication. Resorting to critical multiculturalism might help to resolve this issue, as it involves training the teachers to practice self-reflection and acknowledge their own bias (May, 2009). Gunnþórsdóttir et al. (2018) pointed out that a failure of communication is related to the teachers' lack of training in how to teach immigrant students. It would be beneficial for the teachers to learn about the educational expectations of the parents towards their children, the home language, the values and norms, and how children are taught in their home country. In this study, most of the teachers indicated that they did not receive enough training on how to deal with refugee students and aspired to learn more about the culture and the language. Furthermore, McGee Banks stressed the fact that teachers ought to engage in outreach to the parents, instead of just waiting for them to become involved (McGee Banks, 2010). Even though the teachers interviewed displayed a welcoming attitude, the findings reveal that they did not encourage the parents to be involved more than was necessary. The level of involvement is dependent on the parents' willingness and motivation.

Hamilton, McBrien, and Rah et al. emphasized the importance of assigning same-cultural and bilingual liaisons to enhance parental involvement. Liaisons serve as a cultural bridge between parents, students, and schools and help to develop a clear home–school communication (Hamilton, 2003; McBrien, 2011; Rah et al., 2009). The schools have provided a non-Syrian Arab translator to fill this role. The teachers also pointed out that the municipality assigned an Icelandic woman, Anna, who was responsible for the Syrian families. Anna was referred to when the teachers needed any assistance regarding the refugees, and she too served as a liaison. However, neither the translator nor Anna were from the same cultural background as the Syrians; even though the translator is an Arab, he is not Syrian and does not share a similar history with the Syrian refugees. Additionally, the translator and Anna are the only available liaisons, as the Arab community and, more specifically, the Syrian community, is very small.

Another way to establish a more productive relationship between home and school is to inform the parents about their children's assessment procedure, the teaching techniques, the learning objectives, the required materials and books, and the various ways through which they can provide support to their children's achievement (McGee Banks, 2010). The findings indicate that these factors are missing. In the "Individualized teaching and students' academic achievements" section, it was clear that the parents did not understand how the all-inclusive system works and were not able to evaluate their children's progress. One of the main reasons for the parents' confusion is the language barrier. Teachers explained that the students' learning objectives are written in Icelandic, making it difficult for the parents to understand. Additionally, the individualized assessment procedure, which is based on setting individual goals and assessing the students' level of achievement towards these goals, is new for the teachers. The teachers pointed out they are still "adapting" to this procedure. It was apparent that they did not receive sufficient support on how to implement it adequately, which made it challenging for them to explain it to the parents.

Besides the confusion regarding the inclusive system and assessment procedure, a lack of communication was evident when discussing homework. All of the parents and some of the students demanded homework. Parents complained that their children did not receive enough material to work on at home. Their main reasons for wanting homework were to stay up to date with their children's schoolwork and to be reassured that their children are studying effectively. It was unclear whether the students' reasons for demanding homework were due to their parents' demands or because it is truly what they want. Yet, both the parents' and students' requests contradicted the teachers' responses about homework. The teachers mentioned that they are giving the refugee students homework, but the work is not always completed. In addition, they stated that reading is a daily requirement, and that the students have to complete unfinished classwork at home. According to the teachers, the parents are regularly informed about their children's classwork, either through an email written in English or through the computer software system, which they translate into English specifically for this group. As a result, there is wide ambiguity regarding homework.

Most of the teachers opposed the idea of homework. Even though their opposition stemmed from a moral motive, which is to respect the free time of the students, they were not able to identify the important role homework plays in involving the parents. McGee Banks pointed out that one of the most vital roles parents can play in their children's academic process is working with them. This will "develop a positive self-concept and a positive attitude toward school as well as a better understanding of how their effort affects achievement" (McGee Banks, 2010: 431). Furthermore, simply telling the parents to work with their children is not enough: they need clear directions on how to support their children (McGee Banks, 2010).

Lack of support and training for teachers

Teacher training plays an essential role in creating clear communication channels between home and school, and in decreasing the negative influence of poor cultural fit between two places. Teacher training is an important component when dealing with new refugee children (Frater-Mathieson, 2003). Trainings should be given to enhance the understanding of refugee experiences and trauma (Frater-Mathieson, 2003), and to ameliorate the teacher's knowledge about various cultures including that of the refugees' (Hamilton, 2003). In this study, the teachers revealed that they did not receive enough training on how to deal with refugee families, and did not gain sufficient awareness regarding the Syrian culture and language.

In addition to receiving training around refugee education and the Syrian culture, teachers expressed the stress they experienced owing to the nature of all-inclusive classrooms. Besides the teachers, one of the students highlighted this aspect in the interview, suggesting the school provide an extra teacher in the classroom in order to improve the quality of studying. All-inclusive classrooms are seen as a way to increase social cohesion (Taylor & Sidhu, 2012) and enhance the development of the refugees' host language (Hamilton, 2003). Yet teachers, like students, need extra support, and it is normal for some teachers to feel overwhelmed with additional duties (Hamilton, 2003). Gunnþórsdóttir and Jóhannesson's (2014) study revealed that compulsory school teachers in Iceland perceived inclusive education as an extra workload; additionally, they complained about the lack of resources to target the whole class towards inclusive structures. These results were reflected in the findings of this research. Teachers in this study aspired to have extra support in the classroom and help in preparing material for the refugee students. This issue can be addressed by employing skilled professionals who can work within the system to support regular teachers (Moore, 2003). Finally, teachers complained that there is a lack of psychological support for the students and a lack of specialized personnel to help students with difficulties.

Conclusion and recommendations

Multicultural ideology is considered an effective approach to establishing a successful integration process. In schools, this approach can be achieved through multicultural education. Multicultural education is not merely about enhancing content and reforming the curriculum; it is an entire school reform that targets everyone and everything involved in the schooling process (Banks, 2009). Icelandic schools still have a long way to go to successfully work around and implement multicultural education. Educational policies and national curriculum guides need to be clearer on how to apply multicultural education and how to work with students and families from various cultural backgrounds. Integral parts of multicultural education are involving the parents, establishing clear home–school communication, and properly preparing teachers to deal with diversity. As seen in this study, these elements are missing. In order to enhance home–school communication and increase parental involvement, it is important for teachers and school staff to be trained on how to properly apply multicultural education, be informed about the families' culture, learn their language, and critically practice self-reflection. They need to be active in their approach to get parents involved and provide them with comprehensible information about the school system and their children's education. Offering language and computer training programs can contribute in ameliorating the home–school communication channels. Having skilled personnel, such as psychologists, translators, and bilingual liaisons within the school system can aid the teachers in understanding the needs and difficulties of refugee students and in communicating with the families. Furthermore, providing teachers with suitable materials and in-classroom teaching assistance can ease their stress regarding the nature of all-inclusive classrooms. Finally, in order to better comprehend the acculturation process of the refugee families, it is important to interview them again a few years from now. To understand the refugee students' acculturation in schools in detail, further studies need to be done, such as studies related to their social life and language acquisition. Besides that, similar research with refugee families residing in different locations in Iceland is essential to broaden the understanding of refugee education in Iceland as a whole.

References

Anderson, A. (2003). Issues of migration. In R. Hamilton & D. Moore (Eds.), *Educational interventions for refugee children: Theoretical perspectives and implementing best practices* (pp. 64–82). New York: Routledge.

Banks, J. A. (2009). Multicultural education: Dimensions and paradigms. In J. A. Banks (Ed.), *The Routledge international companion to multicultural education* (pp. 9–32). New York: Routledge.

Banks, J. A. (2010). Multicultural education: Characteristics and goals. In J. A. Banks & C. A. McGee Banks (Eds.), *Multicultural education: Issues and perspectives* (pp. 3–31). Hoboken, NJ: Wiley.

Berry, W. J. (2005). Acculturation: Living successfully in two cultures. *International Journal of Intercultural Relations, 29*(6), 697–712.

Block, K., Cross, S., Riggs, E., & Gibbs, L. (2014). Supporting schools to create an inclusive environment for refugee students. *International Journal of Inclusive Education, 18*(12), 1337–1355. doi:10.1080/13603116.2014.899636

Braun, V., & Clarke, V. (2006). Using thematic analysis in psychology. *Qualitative Research in Psychology, 3*(2), 77–101. doi:10.1191/1478088706qp063oa

Castles, S. (2009). World population movements, diversity, and education. In J. A. Banks (Ed.), *The Routledge international companion to multicultural education* (pp. 49–61). New York: Routledge.

Compulsory School Act No 91/2008. Retrieved October 10, 2018, from www.government. is/media/menntamalaraduneyti-media/media/law-and-regulations/Compulsory-School-Act-No.-91-2008.pdf

Esterberg, K. (2002). *Qualitative methods in social sciences.* New York: McGraw Hill.

EURYDICE. (n.d.). Iceland – Quality assurance. Retrieved November 16, 2018, from https://eacea.ec.europa.eu/national-policies/eurydice/content/quality-assurance-30_en

Ficarra, J. (2017). Comparative international approaches to better understanding and supporting refugee learners. *Issues in Teacher Education, 26*(1), 73–84.

Fife, W. (2005). *Doing fieldwork: Ethnographic methods for research in developing countries and beyond.* New York: Palgrave Macmillan.

Frater-Mathieson, K. (2003). Refugee trauma, loss and grief: Implications for intervention. In R. Hamilton & D. Moore (Eds.), *Educational interventions for refugee children: Theoretical perspectives and implementing best practices* (pp. 12–34). New York: Routledge.

Gunnþórsdóttir, H., Barillé, S., & Meckl, M. (2018). The education of students with immigrant background in Iceland: Parents' and teachers' voices. *Scandinavian Journal of Educational Research.* doi:10.1080/00313831.2017.1415966

Gunnþórsdóttir, H., & Jóhannesson, I. Á. (2014). Additional workload or part of the job? Icelandic teachers' discourse on inclusive education. *International Journal of Inclusive Education, 18*(6), 580–600. doi:10.1080/13603116.2013.802027

Hamilton, R. (2003). Schools, teachers and the education of refugee children. In R. Hamilton & D. Moore (Eds.), *Educational interventions for refugee children: Theoretical perspectives and implementing best practices* (pp. 83–96). New York: Routledge.

Hannah, J. (2007). The role of education and training in the empowerment and inclusion of migrants and refugees. *Education and Society, 25*(1), 5–23. doi:10.7459/es/25.1.02

Hofstede, G. (2011). Dimensionalizing cultures: Hofstede model in context. *Online Readings in Psychology and Culture, 2*, 1. doi:10.9707/2307-0919.1014

Hofstede, G., Hofstede, G. J., & Minkov, M. (2010). *Cultures and organizations: Software for the mind* (3rd ed.). New York: McGraw-Hill.

Hofstede Insights. (n.d.). Compare countries. Retrieved on November 16, 2018, from www.hofstede-insights.com/product/compare-countries/

Joffe, H., & Yardley, L. (2004). Content and thematic analysis. In D. Marks & L. Yardley (Eds.), *Research methods for clinical and health psychology* (pp. 56–68). Thousand Oaks, CA: Sage.

Lerner, B. A. (2012). The educational resettlement of refugee children: Examining several theoretical approaches. *Multicultural Education, 20*(1), 9–14.

May, S. (2009). Critical multiculturalism and education. In J. A. Banks (Ed.), *The Routledge international companion to multicultural education* (pp. 33–48). New York and London: Routledge.

McBrien, J. L. (2005). Educational needs and barriers for refugee students in the United States: A review of the literature. *Review of Educational Research, 75*(3), 329–364.

McBrien, J. L. (2011). The importance of context: Vietnamese, Somali, and Iranian refugee mothers discuss their resettled lives and involvement in their children's schools. *Compare: A Journal of Comparative and International Education, 41*(1), 75–90. doi:10.1080/03057925.2010.523168

McGee Banks, A. C. (2010). Communities, families and educators working together for school improvement. In J. A. Banks & C. A. McGee Banks (Eds.), *Multicultural education: Issues and perspectives* (pp. 417–438). Hoboken, NJ: Wiley.

Ministry of Education, Science and Culture. (2014, March). *The Iceland national curriculum guide for compulsory schools with subject areas.* Reykjavik: Mennta- og menningarmálaráðuneyti. Retrieved November 12, 2018, from www.government.is/publications/reports/report/?newsid=1e0a8c30-ff76-11e7-9425-005056bc530c

Moore, D. (2003). Conceptual and policy issues. In R. Hamilton & D. Moore (Eds.), *Educational interventions for refugee children: Theoretical perspectives and implementing best practices* (pp. 97–105). New York: Routledge.

Nieto, S. (2009). *Language, culture, and teaching: Critical perspectives* (2nd ed.). New York and London: Routledge.

Nieto, S., & Bode, P. (2010). School reform and student learning: A multicultural perspective. In J. A. Banks & C. A. McGee Banks (Eds.), *Multicultural education: Issues and perspectives* (pp. 395–416). Hoboken, NJ: Wiley.

Organisation for Economic Co-operation and Development (OECD). (2017). PISA 2015 results (Vol. III): Students' well-being. Paris: OECD. doi:http://dx.doi.org/10.1787/9789264273856-en

Ragnarsdóttir, H. (Ed.). (2015). Learning spaces for inclusion and social justice: Success stories from immigrant students and school communities in four Nordic countries. Report on main findings from Finland, Iceland, Norway and Sweden. Retrieved on October 12, 2018 from http://lsp2015.hi.is/final_report

Ragnarsdóttir, H., & Rafik Hama, S. (2018). Refugee children in Icelandic schools: Experiences of families and schools. In H. Ragnarsdóttir & S. Lefever (Eds.), *Icelandic Studies on Diversity and Social Justice in Education.* (pp. 82–104). In book series Nordic Studies on Diversity in Education. Cambridge: Cambridge Scholars.

Rah, Y., Choi, S., & SuongThi Nguyen, T. (2009). Building bridges between refugee parents and schools. *International Journal of Leadership in Education, 12*(4), 347–395. doi:10.1080/13603120802609867

Ryan, M. R., & Deci, L. E. (2000). Intrinsic and extrinsic motivation: Classical definitions and new directions. *Contemporary Educational Psychology, 25*(1), 54–67. doi:10.1006/ceps.1999.1020

Sheikh, M., & Anderson, R. J. (2018). Acculturation pattern and education of refugees and asylum seekers: A systematic literature review. *Learning and Individual Differences, 67*(1), 22–32. doi:10.1016/j.lindif.2018.07.003

Taylor, S., & Sidhu, K. R. (2012). Supporting refugee students in schools: What constitutes inclusive education? *International Journal of Inclusive Education, 16*(1), 39–56. doi:10.1080/13603110903560085

Thomas, R. L. (2016, December 7). The right to quality education for refugee children through social inclusion. *Journal of. Human Rights and Social Work, 1*(1), 193–201. doi: 10.1007/s41134-016-0022-z

Topping, K., & Maloney, S. (2005). Introduction. In K. Topping & S. Maloney (Eds.), *The RoutledgeFalmer reader in inclusive education* (pp. 1–14). New York: Routledge.

Vaismoradi, M., Turunen, H., & Bondas, T. (2013). Content analysis and thematic analysis: Implications for conducting a qualitative descriptive study. *Nursing and Health Sciences, 15*(1), 398–405. doi:10.1111/nhs.12048

4 Immigrant youth perspectives

Understanding challenges and opportunities in Finnish Lapland

Ria-Maria Adams

The impacts of global migration are increasingly visible in the Arctic regions. Immigration is under constant change, and questions of diversity and integration are core issues addressed by social sciences and policymakers. Human migration is an age-old phenomenon that can be traced back to the earliest periods of human history. However, this era of mass emigration and immigration provides states and societies with new challenges and opportunities (Autto & Nygard, 2015; Jokela & Coutts, 2017; Petersen & Poppel, 1999). The United Nations' International Migration Report 2018 states that the estimate for the year 2015 was that 244 million people were residing outside their country of citizenship. The latest projection for the year 2050 is a global rise in migration numbers to 405 million. The overwhelming majority of people migrate for reasons related to work, family, and study. Concurrently, many people leave their home countries for other compelling reasons, such as conflict, persecution, and disaster (McAuliffe & Ruhs, 2017). International migration is expected to increase owing to several factors such as population increase, economic conditions, structurally caused underdevelopment, political conflicts, or ecological factors (Martin, 2015). South to North migration flows are being analysed and interpreted within different scientific disciplines, as they have become a global phenomenon (Dimova- Cookson & Stirk, 2010; Fortier, 2000; Kokot & Dracklé, 1996; Moore, 2009; Vertovec, 2017).

Nordic countries have seen an increase in immigration, especially in 2016, which reached a historic level of immigration in the region (Heleniak, 2018). The migration policies of Nordic countries have long been welcoming towards migrant workers, and the region continues to be an attractive place for newcomers because of its strong economics. Immigrants are especially important to rural regions, such as Finnish Lapland, with declining population numbers (Heleniak, 2018, p. 48). The Second Arctic Human Development Report (Larsen & Fondahl, 2014) summarizes the current state of research in a broad sense by addressing regional processes and global linkages, and the Arctic Social Indicators II report (Larsen et al., 2014) focuses on tracking the change in human development in the Arctic. Rapid change is visible in the socio-economic transformations of daily living, which have led to the need to rethink human well-being and community adaptability in the region (Larsen et al., 2014, p. 15; Petrov et al., 2017). Migration has become the major source of population increase in the European Nordic countries; between the years 2000 and 2018, the population of Denmark, Finland, Iceland, Norway, and Sweden increased by 2.7 million (Heleniak, 2018, p. 48). Integrating immigrants into the Nordic countries has become both a challenge and a priority for policymakers. The Finnish government is also concerned with the development of an inclusive immigration policy that integrates strategies to meet both labour and human needs. In the Nordic countries, extremes such as long winters, sparse and declining populations, as well as cultural adaptation are a challenge. Integration is especially visible in education policy, where inclusive measures are being promoted (Angeria & Niskanen, 2019; Heikkilä & Peltonen, 2002; Taskiainen et al., 2019; Yeasmin, 2012).

The official Finnish government report (TEM, 2019) shows that the total number of inhabitants in Finnish Lapland in 2017 was 179,223. The population of Finnish Lapland consists mainly of Finnish people, but also includes indigenous Sámi people (Määttä & Uusiautti, 2019, p. 215), with a population density of only two persons per square kilometre (Petäjämaa, 2013). Only 20 per cent of Lapland's inhabitants live in rural areas, and the remaining 80 per cent reside in the few towns of the area (Petäjämaa, 2013, p. 4). Although the overall population in the area has been constantly decreasing, the population of immigrants has been steadily increasing. In the year 2017, 4013 immigrants lived in Finnish Lapland, 2.2 per cent of the total population. The government is implementing new strategies in this area to meet its intentions to increase the number of inhabitants through economic growth, raising birth rates, migration, immigration, and tourism, by developing the mining sector. The prediction is that, in the year 2030, 12,000 immigrants would be living in the northern areas of Finland, most of them in the city of Rovaniemi. This would require an annual increase of 7 per cent, which is rather optimistic, as immigration in the year 2017 only grew by 0.5 per cent (TEM, 2019). The current immigrant population speaks around 80 different languages, most commonly Russian, Arabic, English, and Thai, which are also the immigrant youth's native languages (TEM, 2019). Consequently, the number of students from an immigrant background in Finnish Lapland is low. In the city of Rovaniemi, which is the largest city in the area, approximately 190 students were categorized as potential participants for Taskiainen et al.'s (2019, p. 145) study on enhancing immigrant students' participation in Arctic schools in 2018, and my current research operates with the same numbers. Adding immigrant youth from all over Finnish Lapland would barely double the amount. There is a lack understanding of youth immigrants' cultural well-being in Finnish Lapland, partly as it is a relatively new development. In 2017, 1346 immigrant youth aged between 15 and 29 years were residing mostly in the bigger cities of Rovaniemi and Kemi. By comparison, the number of young immigrants in the same age range in the year 1990 was only 118 (State Youth Council, 2019). However, government surveys can be insufficient for understanding of the complex process of cultural adaptation undergone by young immigrants in Finland. Hence, there is a need to add more knowledge to the discourse so that the youths' own voices will be more audible. Therefore, the leading research question for my ongoing study is: What are the opportunities and challenges that young immigrants face while living in

Finnish Lapland? Whereas the Finnish government's endeavours are concerned with vitalizing the sparsely inhabited regions of the country, young people are more preoccupied with making the decision to stay or to leave the region. The underlying reasons behind such a decision, or feeling connected or unconnected to Finnish Lapland, are closely linked to different factors of well-being that will be discussed from both theoretical and empirical perspectives in this chapter.

Theorizing migration has grown to be a central focus within the anthropological field in recent years (Brettell, 2003; Eriksen, 2009, 2010 [1995]; Fortier, 2000; Moore, 2009). However, there is a lack of examination of the specific field of youth migration within the conceptual limits of well-being and cultural adaptation, especially in the regional context of the Arctic. Steven Vertovec's (2011) view on diversity serves as an example of how the term "diversity" is often gratuitously used on a policy level in migration-related issues, without a clear understanding of the actual meaning of the term. Vertovec addresses the issue of immigrant cultures routinely being posed as threats to national cultures. Moreover, he states that there is a particular understanding of immigrant and national cultures, which are mutually constituted in policies and state institutions, the media, and everyday perceptions surrounding key categories such as borders, illegality, and the law (Vertovec, 2011, p. 241). Therefore, it is essential to examine how young immigrants are being represented in research and studies, and how terms such as "diversity" or "multiculturalism" are used in discussions on issues related to their own perceptions of life. The use of an anthropological lens in analysing this topic provides a holistic perspective that does not rely only on the substance of qualitative interviews, but goes further by participating and engaging in the daily life and activities of immigrant youth. Interest in a successful integration of the young newcomers to Finnish Lapland is shared by several stakeholders, such as the European Union, the Arctic research communities, local governments, and economists who are searching for innovative solutions to the occurring phenomena. Young people who are going to school, learning a new language, managing multiple identities, and longing for peer groups must be better understood in their development.

This chapter argues that an anthropological lens provides an important understanding of young immigrants' perspectives on life in the Arctic, more specifically focusing on immigrant youth perspectives in Finnish Lapland. First, a literature review of theories on diversity, multiculturalism, and factors contributing to immigrant youth well-being serves as the theoretical basis of the chapter. Next, empirical data from my own research in Finnish Lapland is provided to discuss the perspectives of young immigrants in viewing threats to, and opportunities for, their own well-being. The chapter concludes by discussing the implications surrounding future research on the topic of immigrant youth in Finnish Lapland.

Through the lens of anthropology

Anthropology is about understanding humans' point of view and how they relate to life. Olwig and Hastrup (1997, p. 1) argue the idea that cultures can be conceptualized as separate and unique entities corresponding to particular localities, which has been a means of erecting frameworks for perceiving and understanding cultural differences. Brettell and Hollifield (2015) provide an explanation as to how the study of migration could be constructed:

> Anthropologists who study migration are interested in more than the who, when, and why; they want to capture through their ethnography the experience of being an immigrant and the meaning, to the migrants themselves, of the social and cultural changes that result from leaving one context and entering another. [...] Questions in the anthropological study of migration are framed by assumption that outcomes for people who move are shaped by their social, cultural, and gendered locations and that migrants themselves are agents in their behaviour, always interpreting, constructing, and reconstruction social realities within the constraints of structure.
>
> (Brettell & Hollifield, 2015, p. 5)

The role of anthropological research has expanded from engaging with local to global contexts (Crate, 2011; Eriksen, 2010 [1995]). Anthropological research is valuable because of its holistic approach, taking all different aspects of human living into account and considering both actions and verbal accounts. Anthropology is also known as the discipline devoted to understanding and dealing with cultural difference. Crate summarizes the essence of anthropological work as follows:

> We work on many scales, from local to global, wear many different "hats", from that of academic researcher to advocate, engage issues of culture by interpreting both cultural breakdown and culturally based responses, and continue to use some of anthropology's foundational methodologies including ethnography, participant observation, interpretation, documentation, and the like but within a newer global context of issues, collaborators, and audiences.
>
> (Crate, 2011, p. 179)

Moreover, the anthropological quest is to investigate human diversity and not human homogeneity (Eriksen, 2017, p. 1147) in all its facets. Research based on anthropological methods of fieldwork tends to add a deeper understanding of immigrant youth. Anthropologists have contributed to theories on migration, mostly in local and global understandings of how migration movements are affecting the world (Brettell, 2003; Vertovec, 2009, 2017). However, according to Crate (2011), anthropologists should push their boundaries further and become more globalized agents for change. She suggests that anthropologists become more active as public servants and take active, participatory roles in designing policy guidelines. Furthermore, she emphasizes that anthropology's task is to connect research and knowledge to those who are actively involved in emerging issues to facilitate global understanding and scope (Crate, 2011, p. 184). In the context of my own research, this means giving a voice to young immigrants, letting their own view of immigrant experiences be heard, and building a bridge between the actors (immigrant youth), science, and policy. Anthropology has a unique capacity to identify, track, describe, and interpret in depth, but it also carries responsibility for passing knowledge on to others. Correspondingly, it is the task of anthropologists not to judge the way in which immigrant youth are constructing their lives, but to make sense of their world in the way they perceive it. Tsuda et al. (2014) suggest that traditional ethnography, which is based on in-depth fieldwork in one specific place, is sufficient to capture the intricacies of an increasingly globalized world and shows the transnational connections between various localities.

To understand the perspective of immigrant youth, their current challenges, and opportunities, anthropological fieldwork methods, which include participant observation and in-depth interviews, are functioning and efficient tools (Bernard, 2000; Clark, 2011; Emerson et al., 1995; Flick, 2014 [1998]; Fraser et al., 2004; Silverman, 2013, 2014). This current research includes first-generation immigrant youth, aged between 13 and 18 and from all nationalities, with a permanent residence permit in Finland. Because of the young age of the research participants, specific child-centred, qualitative research methods were used (Berman, 2011; Clark, 2011; Delgado, 2006; Fraser et al., 2004; Lancy, 2008), which favour an age-appropriate approach and value listening to the younger voice in a way that "treats them not as human becomings, but human beings" (Clark, 2011, p. 16). Kehily (2007) states that anthropological approaches about youth suggest that, during this transnational period in their lives, young people carry out various roles and meanings in different cultures. The value of ethnographic research practices emerges when the viewpoints of the researched group are included. It is both essential to understand the general framework in which immigration is taking place in Finnish Lapland, as well as examine specific local contexts. In practical terms, this research method requires spending sufficient time in the field and connecting with the immigrant youth in their own spaces, which are: schools, public spaces in city centres, shopping mall(s), youth centres, sports clubs, government-funded immigrant integration projects, and homes. Active engagement in their fields of interest is necessary when it comes to applying this method in a useful manner. For the past 2 years, research has mostly been conducted in the city of Rovaniemi. Concomitantly, examples from the city of Kemijärvi and the municipality of Kolari are included in the study. As a result, youth perspectives throughout the region provide new insights to the development of sustainable communities in Finnish Lapland.

Immigrant youth well-being and diversity

How are diversity and well-being constructed for immigrant youth in Finnish Lapland, and how do they deal with it in their daily lives? How do these aspects exist in Finnish Lapland, and what are the challenges for youth in this particular Arctic region? The Finnish Executive Summary of the Government's Report on the Implementation of the 2030 Agenda for Sustainable Development has defined as one of their main focus areas a "non-discriminating, equal and competent

Finland". They especially focus on the well-being of children and youth in a time when inequality has increased in Finland (Finnish Government Report, 2018) and state:

> According to the values espoused by Finnish society, everyone is a valuable and an equal member of the society, and everyone has equal opportunities for wellbeing and good life, health, working and functional capacity, education and employment. Everyone is entitled to participate in the society in a meaningful way. Ensuring wellbeing of children and young people, and supporting civil engagement are of particular importance. Promoting gender equality is an integral part of the implementation of all goals of the 2030 Agenda.
>
> (Finnish Government Report, 2018, p. 7)

This is valid for immigrant youth in the same way as it for their Finnish counterparts. But how is this well-phrased and well-intentioned policy being implemented on a practical level? Well-being is addressed in various policies aimed at young people, and the scientific task within this topic is to explain the meaning of well-being and how it is experienced individually and collectively. Once the cultural specificities are understood, assumptions on a broader human level can be made. Mathews and Izquierdo (2009) raise the topic of happiness and well-being in anthropology and social sciences because of its frequent use by academics, politicians, and NGOs. According to Dimitrova (2017), well-being is one of the most commonly studied psychosocial subjects among children, youth, and adults, but there is a lack of studies regarding the phenomenon of well-being in a broader, societal context. The precise meaning of well-being is elusive and ambiguous, but the term is generally used in psychology to refer to indices of positive adjustments, and a flourishing and thriving physical and psychological state (Dimitrova, 2017, p. 8). Bobowik et al. (2017) state that the well-being of ethnic minority youth has been mostly measured in negative terms such as depressive mood, poorer mental or physical health, or distress (Bobowik et al., 2017, p. 158). However, the cultural differences in perceiving well-being on an individual and collective level, and the interpretations of well-being, vary between different cultures. (Dimitrova, 2017, pp. 8–9). Hence, my research aims to examine the similarities that immigrant youth share in their experiences in Finnish Lapland. Are there aspects of cultural adaptation and well-being that are collectively experienced? I will later argue that there are aspects, such as education, security, infrastructure issues, and the feeling of belongingness, that are shared among young people, regardless of their origin.

Minors are still rarely studied as active social agents with their own rights, needs, and desires. Matthews et al. (1999, p. 135) argue that, unlike other marginalized groups, children are not in a position within most societies to enter into a dialogue about their environmental concerns and geographical needs, and, in this sense, young people occupy a special position of exclusion. According to the European Commission Youth Policies Report (2017), Finland must be acknowledged as one of the leading countries with its own laws for respecting the rights of young people. The Finnish Youth Act, for example, promotes social inclusion and opportunities to participate in policy- and decision-making. The development of individuals' abilities is stipulated, and free-time hobbies and youth work are available in the small municipalities of the North. The government supports the implementation of the Act on national, regional, and local levels. However, it is still difficult to implement practices to incorporate immigrant youth into various activities that promote their inclusion within mainstream society. While Bobowik et al. (2017) explain that social integration refers to belongingness, social acceptance includes an accepting view of human nature and trust in others. Christ et al. (2014) found in their psychological study that positive face-to-face contact, especially between cross-group friends, has a positive effect on individual well-being. Supposedly, contact between different ethnic groups reduces prejudice among minority and majority group members (Christ et al., 2014, p. 3996). Kehily (2007) suggests that youth can be understood and conceptualized in different ways, and she brings the cultural viewpoint into the debate. She advocates for the need to distinguish between looking at youth from cultural, comparative, and biographical perspectives. The cultural aspect is defined by Kehily as "everyday social practice that young people make sense of the world and take their place in it through participation and engagement with the routine social practices of everyday life" (Kehily, 2007, p. 4). Promoting the well-being of children and youth will have societal effects (Pollock et al., 2018, p. 1), and therefore it is important to include this in the policy agendas of local governments, EU institutions, and UN declarations.

Prevailing migration theories deal with reasons for migration and explain the current situation of the movement of people (Bhabha, 2014; Brettell, 2003; Castles & Miller, 2009; Ensor & Gozdziak, 2010; Hammar et al., 1997; Hirvi, 2013; Koser, 2016; Vertovec, 2010). Caglar and Glick Schiller (2018, p. 5) argue for a multiscalar analysis beyond methodological nationalism and the ethnic lens, and appeal for migrants to be approached as social actors who are integral to city-making on various different levels, which becomes apparent especially in the study of young immigrants. Discourses on the cultural adaptation of youth have been discussed in contexts that often explain exclusive, criminal, educational, or traumatized subject matters (Bhabha, 2014; Ensor & Gozdziak, 2010). As the terms "diversity" and "multiculturalism" appear within the subject of immigrant youth well-being, it is vital to engage with this topic from a theoretical viewpoint. Vertovec's concepts of diversity and multiculturalism provide a good introduction and perspective on addressing these commonly used, but rather vague, terms.

Vertovec (2001) has always advocated "diversity" as a descriptive concept and approach to new migration patterns, but at the same time he emphasizes that it is not a theory (Meissner & Vertovec, 2015; Vertovec, 2001, 2017, 2019). A theory would regard inherent relations or causalities, and a hypothesis would be a "to-be-tested theory", which the concept of "super-diversity" is not. Vertovec (2019) came up with the term "super-diversity" in order to explain new migration patterns of inequality and prejudice:

> Super-diversity is a summary term proposed also to point out that the new migration patterns not only entailed variable combinations of these traits, but that their combinations produced new hierarchical social positions, statuses or

stratifications. These, in turn, entail: new patterns of inequality and prejudice including emergent forms of racism, new patterns of segregation, new experiences of space and "contact", new forms of cosmopolitanism and creolization.

(Vertovec, 2019, p. 126)

Vertovec provides an explanation for how and why these changing patterns arose, as well as their interlinkages and consequences (Vertovec, 2017, 2019). He also makes a clear statement that super-diversity does not merely mean more ethnicity, meaning that new migration processes have brought more ethnic groups to a nation or city (Vertovec, 2019, p. 130). Consequently, he advocates for the need to rethink approaches used within the social sciences to deal with migration issues, and to move beyond an "ethno focal lens", which is still used within conventional migration studies (Meissner & Vertovec, 2015, p. 542). Concomitantly, Vertovec addresses diversity at a policy level (Meissner & Vertovec, 2015; Vertovec, 2012) and states that "'diversity' is the focus of a wide-ranging corpus of normative discourses, institutional structures, policies, and practices in business, public sector agencies, the military, universities, and professions" (Vertovec, 2012, p. 287). Diversity is also the key word in making EU youth plans, diversity programmes, and training strategies, and the Finnish Youth Act aims for integrative measures (European Commission, 2017). Vertovec (2012) argues that the language of "multiculturalism" has largely been changed into "diversity". Furthermore, he states that the "European Charters for Diversity" are made for organizations and businesses with efforts to promote "diversity" and combat discrimination (Vertovec, 2012, p. 294), which is visible in the Finnish education system as well. Moreover, the diversity discourse arose and developed for various reasons, ranging from addressing historical disadvantages and discrimination all the way to avoiding the risk of discrimination lawsuits and to strategies for gaining benefits for organizations (Vertovec, 2012, pp. 294–295). Through multivalence and optimistic orientation, diversity has become an ever-present emblem of fairness and openness (Vertovec, 2012, p. 302).

Despite this omnipresence and formal implementation of diversity, practical inclusion is lacking on various levels, as I will argue later on through empirical examples. Therefore, it is not enough to just address diversity for the sake of "political correctness"; there is a need to more practically address diversity in Arctic youth issues. Ironically, according to Vertovec, most people are not sure what "diversity" refers to, and yet the term has a positive connotation for most (Vertovec, 2012, p. 307). The preliminary results of my current research among immigrant youth coincides with Vertovec's perception of the terminology use: in schools or youth centres, diversity strategies and policies are incorporated theoretically in the strategies and plans, but the practical implementation of getting young people from different backgrounds to interact with each other is a more complex endeavour. The effort of including a "newcomer" requires more effort than tolerating immigrants simply being present. Vertovec reminds us that, for more than 30 years, international policies have had as their overall goal the promotion of tolerance and respect for other groups and identities. This task has modified public services by promoting culture-based differences of value, language, and social practice (Vertovec, 2010, p. 83). Moreover, Vertovec (2017) points out that, at a practical level, every state has introduced policies regarding opportunities, restrictions, and consequences for immigrants. These range from accessing welfare, health, education, work, public services, justice system, length and nature of residence, to chances of permanent settlement and the possibility of gaining citizenship (Vertovec, 2017, p. 1576). However, he also reminds us that migrants adapt in one way or the other:

> The incontestable fact is that with regard to both transnationalism and integration, migrants adapt. Sustained and intensive patterns of transnational communication, affiliation and exchange can profoundly affect the manner of migrant adaptation – including practices associated with positive or limited integration – through the maintenance of a particularly strong sense of connection or orientation to the people, places and senses of belonging associated with the place of origin.
>
> (Vertovec, 2010, p. 99)

What Vertovec (2019) proposes is a new approach to new migration patterns with multidimensional categories, shifting configurations, and new social structures, which is relevant to Arctic immigration. There is a need to focus beyond ethnicity, as ethnic groups are not the optimal units of analysis and may actually mask more significant forms of differentiation (Vertovec, 2019, p. 131). Addressing diversity or super-diversity in the context of Finnish Lapland might, at first glance, seem incomprehensible, given the overall low number of immigrants. However, Vertovec does not address the density of immigrants at any time, but builds his argument on the conceptual limits of diversity. In that sense, the theoretical engagement with terms such as "diversity" or "multiculturalism" is not limited to places with large immigration numbers, but can be used in any context where topics of diversity and multiculturalism are being discussed. They provide ideas for understanding the embeddedness of immigrant policy strategies, which are interpreted in this research through the different factors that impact the well-being of immigrant youth.

Results: opportunities and challenges for young immigrants

The selection criteria included first-generation male and female immigrant youth, aged between 13 and 18, from all nationalities with a permanent residence permit in Finland. They are junior high school (*yläkoulu*), high school (*lukio*), and vocational training (*ammattikoulu*) students. The majority of these young people have not chosen this Arctic region as their new home but have moved along with their families. Suárez-Orozco et al. (2011) describe migration as the human face of globalization and remind us that where immigrant workers are summoned, their families and children are most likely to follow. The ethical standards of good scientific practice have been met by having consent forms detailing the study signed by both the research participants and their legal guardians.

The results of this study reveal that young immigrants value the Finnish education system. No matter what their country of origin is, and despite their relatively young age, they all agree that the level of education in Finland is excellent and is significant for their personal future career development. The majority of the immigrant youth interviewed speak respectfully about their teachers and perceive them as helpful agents in their personal cultural adaptation process. Therefore, the influential role of teachers and other school personnel, such as social workers and student counsellors, should be recognized as a key factor in integration. Finland's teacher education is highly valued, appreciated, and recognized on an international level, which is reflected in the educational environment of children and youth (Määttä & Uusiautti, 2019, p. 215). Young immigrants also value the opportunities for equal possibilities in education and are enjoying free, well-organized education. Learning the language is viewed as an important part of integration, although some students have difficulties with learning the Finnish language. They often communicate in other languages with their peers if they have the opportunity to do so. Yeasmin (2012) states that the Finnish government generally welcomes students from all over the world to study in various educational institutions, which is perceived and considered positive. Her study reveals that attitudes towards student immigrants are more liberal than towards other immigrants. According to policies, immigrant children and youth are treated in the same way as Finnish children by the Finnish authorities (Yeasmin, 2012, pp. 343–344; Taskiainen et al., 2019). However, Vertovec (2012) notes that simple mentioning of diversity in policies does not automatically imply a practical implementation in everyday life. Children and young people spend most of their active time during the week at educational institutes, and, therefore, a special focus needs to be put on schools when addressing integrative measures (Uusiautti & Yeasmin, 2019). Taskiainen et al. (2019, p. 144) argue that participation is important to immigrant children and youth, especially in learning the host country's language. This results in positive attitudes towards immigrants and helps integrate them into their host country. Nevertheless, being able to communicate in the same language is not enough to be integrated into the local community. There is a deep desire to be part of social and peer groups beyond the school buildings, which will be discussed in the following section on challenges. However, the results of this study show that, for some immigrant youth, the only reason to stay in Finnish Lapland is the outstanding standard of education. The Finnish education system is perceived as a major opportunity for young immigrants, and the extent to which getting a Finnish education can serve as a major pull factor for this area should not be underestimated. Nevertheless, as soon as the compulsory level of education has been reached, the tendency to move elsewhere is obvious, given the limited options of university and vocational studies in the area. The reasons that immigrant youth give for needing to move elsewhere to achieve a higher level of education are identical to those given by Finnish youth, which have been discussed in the Arctic Centre's ongoing project "Live, Work or Leave? Youth – wellbeing and the viability of (post) extractive Arctic industrial cities in Finland and Russia" (Stammler & Adams, 2019).

Another reason that young immigrants give for staying in Finnish Lapland is general safety. The research results show that young immigrants view Finnish Lapland as an extremely safe place in which to grow up and move around. Youth who have moved to this Arctic region, especially those moving from larger urban cities, value the freedom they have in moving around Arctic towns on their own, without fearing crime. The harsh weather conditions limit their mobility outdoors, but nevertheless they express the feeling of not having to be afraid when moving between places on their own. The descriptions of how free they feel when, for example, they are able to cycle alone, with no fear of traffic, are remarkable. Hence, safety issues can also serve as pull factors when opportunities for young immigrants in the Arctic region are assessed. Ultimately, the larger question remains as to whether education and general safety provide enough reasons to stay in Finnish Lapland. The short answer would be no. There are other factors of youth well-being, such as meaningful relationships, a satisfactory choice of spare-time activities and entertainment possibilities, sufficient medical care, perspectives of future work possibilities, and a functioning infrastructure that are also important considerations. It is the combination of many aspects that influences the overall well-being of youth, and the bare provision of excellent education or safety is insufficient for making immigrant youth stay in the long run. There is a high probability that immigrant youth, like their Finnish counterparts, will move away from the area.

Next, I will shift to focus on some of the challenges addressed by youth immigrants. One of the continuously solicited topics is the lack of public infrastructure in the area. Kuhmonen et al. (2016) argued that the stagnant and old-fashioned image of remote rural areas and the countryside represented in the media and tourist advertisements repulses youth, while simultaneously attracting tourists. Although all airports in Lapland are declaring increasing numbers of passengers – in the year 2018, more than 1.3 million passengers travelled to northern destinations (Heikkinen, 2019, p. 34), the infrastructure for people living in the northern regions is not changing as rapidly. On the contrary, given the region's low population density, there are several economic challenges at hand. Although tourists may be willing to pay up to 500 euros per night for a small cabin with glass ceilings to spot the Northern lights (Heikkinen, 2019, p. 34), local and immigrant youth would rather see development in the growth of an accessible, affordable, regular public transportation system within the city limits. Finnish Lapland has the space and place for many more migrants but, without investing in the current infrastructure of such remote places, it will not attract young people, locals, or immigrants to stay and build a new life. Heikkilä (2019) describes Arctic visions, the origins and the background of Arctic cooperation between 1998 and 2018, and states that this northern region would be a "strategic reserve" for the future of Europe. Given the impact of climate change, the planet's forests are disappearing, and clean water is becoming a scarce commodity. The northern regions could secure the energy management of Europe in the next millennia (Heikkilä, 2019, p. 8). Furthermore, he addresses the issue of concrete infrastructure projects planned over time, which were promised but are yet to be implemented (Heikkilä, 2019, p. 12). Typically, the distances between northern communities are great, and the immobility of young people and those who are dependent on public transportation leads to frustration. One effect of insufficient infrastructure is isolation, which for most young people is unbearable over continuous periods of time and further results in a desire to move elsewhere. According to the preliminary results of the ongoing Arctic Centre's youth project, "Live, Work or Leave? Youth – wellbeing and the viability of (post) extractive Arctic industrial cities in Finland and Russia" (Stammler & Adams, 2019), young people dream about having work that pays well and living close to or in a city, with the possibility to interact with people and the surrounding nature. These research results suggest that, in the case of infrastructure and mobility, the same aspirations are shared by both Finnish and immigrant youth. Hence, this implies that immigrant youth have motivations overlapping with those of their Finnish counterparts for moving away from Finnish Lapland: if there are no possibilities for personal mobility and a working infrastructure, the desire to go elsewhere is obvious to them. Kuhmonen et al. (2016) note:

> The Finnish youth barometer observes that large cities attract young people more than rural locations especially due to opportunities for employment and education. Part of the youth considers the urban fabric also more tolerant and liberal than the rural habitat.
>
> (Kuhmonen et al., 2016, p. 91)

Moreover, the futures of local residents and development of rural areas are linked by multidimensional welfare factors, such as the availability of diverse employment, housing, and lifestyle opportunities (Kuhmonen et al., 2016, p. 97). These lifestyle opportunities refer, in the case of my research, to having access to various hobbies, spare-time activities, restaurants, cafes, and shopping malls. The lack of "things to do" in Finnish Lapland, combined with an insufficient infrastructure, results in an undefined wish for "something else". This illustrates how the global effects of consumption are interwoven in the perception of young people, connecting personal well-being to having access to mobility and activities.

On the other hand, the results also indicate that social inclusion and a feeling of belongingness are key in adaptation and building a connection with a place. For immigrant youth, one of the biggest challenges is to find social connections and to find their place within the majority society. Although attitudes of immigrant youth vary on an individual basis, some longing for more social connections and some seemingly content in their position as an "outsider", the common perception is the wish for more connections with their Finnish counterparts. Most young people describe experiencing their Finnish peers as reserved, introverted, and "difficult to approach", despite trying to actively engage with them. First and foremost, it takes a lot of time and patience to get one's bearings in this area of the Arctic. The active engagement of immigrant youth with local youth activities, such as sports and hobby clubs, political youth organizations, religious organizations, and youth centres, is beneficial in the overall integration process, because these provide a "natural environment" to spend time with peers. Nonetheless, it is common that immigrant youth find their closest friends among other immigrants as they share the experience of "being different", and often the shared experience of being excluded connects them even more closely. The social exclusion is often not intentionally caused by the Finnish youth, but appears especially in activities that are typical of the region. In the winter, for example, some Finnish youth (especially males) have access to snowmobiles, which they use for excursions in the area; in the summer, they drive in circles around towns on their mopeds. In some cases, the parents might view these activities as too dangerous or redundant, others have indicated that they simply do not like to do activities outdoors during the harsh winter, and still others do not have the financial resources to participate in such activities. Social exclusion can be traced back to many individual traits, but still there is also a collective longing for social connection with their Finnish peers. Digitalization has certainly changed connectivity a lot in recent years, and Bobowik et al. (2017) argue that the social support in networks and collective efficacy is particularly important for young immigrants' well-being (Bobowik et al., 2017, p. 158). Further examination is needed of how social media and digital connectivity can help with the challenge of integration and cultural adaptation in this northern region.

By virtue of the topics discussed above, Finnish Lapland has enormous potential to offer young immigrants in terms of high-quality education and safety. Although educational institutes and safety aspects play an important role in enhancing youth well-being, these are not the only factors that can lead to a more sustainable Arctic region. Other aspects, such as infrastructure and social integration, should be further developed if the aim is to keep young immigrants in the area.

Conclusion and discussion

In this chapter, I have demonstrated that the perception of young immigrants is that Finnish Lapland is a safe environment to grow up in, and that the educational sector bears enormous potential and opportunities. On the other hand, the lack of sufficient infrastructure and work opportunities is viewed as a reason to leave the area. Social inclusion within broader society matters in terms of integration and well-being, and immigrant youth perceive that it is often hard to connect with their Finnish peers. Pursuing anthropological approaches matters substantially in the research on immigrant youth as it allows a broad view on various topics, emerging from young people themselves. The well-being of immigrant youth is reflected in their own stories and everyday actions and can best be revealed through participating and engaging in their everyday lives. Vertovec's approach to diversity and multiculturalism shows that the concept of diversity should not just be added for the sake of political correctness, but should be genuinely implemented in practice.

Heikkilä quotes the Finnish president Sauli Niinistö, "If we lose the Arctic, we lose the whole world" (Heikkilä, 2019, p. 139), to demonstrate the urgency of Arctic matters. Currently, the Arctic region is perceived as a safe place to live by local and immigrant youth, but what does the future look like? Heikkilä warns us that the Arctic can be lost in several different ways: for example, through global warming, but also through an increasingly polarizing political climate. He argues that both involve changes that happen in small steps until everything has irreversibly changed (Heikkilä, 2019, p. 140). Simultaneously, people's awareness of social and environmental issues is growing, and there is an increased desire among people to engage and interact with such issues (Eriksen, 2017). Suárez-Orozco et al. (2011) remind us that immigration will profoundly change the societies in which immigrants settle, and that the children of immigrants are set to reshape the future character of the world, which implies that our future is their future. Furthermore, the transition of immigrant-origin children to immigrant-dependent countries is a topic of scholarly interest, relevant in policy-making, and practical urgency (Suárez-Orozco et al., 2011, p. 313). Määttä and Uusiautti (2019) emphasize that an increase in immigrants in Finland should be seen as an opportunity to support the Arctic region, and they emphasize Arctic pedagogy as a means to create new kinds of learning and teaching models. Kuhmonen et al. (2016) remind us that those who are committed to stay seem to have a realistic-optimistic vision, whereas others sense some "greener grass" outside the region:

> Evidently, push and pull factors differ among these areas and so does their framing, importance and impact among the youth. It is important to know about the images, dispositions and preferences of the youth, since they affect actual migration behaviour.
>
> (Kuhmonen et al., 2016, p. 91)

Yet, the voice of youth often remains unheard in public policy-making. This chapter has attempted to show that adding the voice of youth and their viewpoints could change the direction of policy-making by yielding unexpected viewpoints. Young people are waiting to be heard. Adding their insights into research and policy-making processes, rather than using "diversity" as an empty phrase, will have an impact on their own futures. Therefore, focusing on factors promoting the well-being of young immigrants and listening to their voices should be a priority when making policies and conducting research, as these actions will consequently lead to better solutions and impacts for sustaining both Finnish Lapland and the broader Arctic region.

Acknowledgements

This chapter was supported by the University of Vienna, funded by the uni:docs fellowship programme.

This chapter was partially supported by the WOLLIE project, Arctic Centre, Anthropology Research team, financed by the Finnish Academy, decision number 314471.

References

Angeria, K. & Niskanen, M. (2019). *Osallisuuden Polku Maahanmuuttajanuorille: Käsikirja Nuorten Parissa Työskenteleville* [Participation Path for Youth Immigrants: Handbook for Youth Workers]. Helsinki, Finland: Diakonissalaitos. Retrieved from: https://s3-eu-central-1.amazonaws.com/evermade-hdl/wp-content/uploads/2019/06/20122433/HDL_Osallisuuden-polku-maahanmuuttajanuorille_sivuttain.pdf

Autto, J. & Nygard, M. (2015). *Hyvinvointivaltion Kulttuurintutkimus* [Cultural Study of a Welfare State]. Rovaniemi, Finland: Lapland University Press.

Berman, E. (2011). The Irony of Immaturity: K'iche' Children as Mediators and Buffers in Adult Social Interactions. *Journal of Childhood*, 18(2), 274–288.

Bernard, R. H. (2000). *Handbook of Methods in Cultural Anthropology*. Lanham, MD: AltaMira Press.

Bhabha, J. (2014). *Child Migration & Human Rights in a Global Age*. Oxford: Princeton University Press.

Bobowik, M., Basabe, N., & Wlodarczyk, A. (2017). Only Real When Shared: Social Well-Being, Collective Efficacy, and Social Network among Immigrant Emerging Adults in Spain. In Dimitrova, R. (Ed.) *Well-Being of Youth and Emerging Adults Across Cultures. Novel Approaches and Findings from Europe, Asia, Africa and America* (pp. 157–171). Cham: Springer.

Brettell, C. B. (2003). *Anthropology and Migration. Essays on Transnationalism, Ethnicity, and Identity*. Oxford: AltaMira Press.

Brettell, C. B. & Hollifield, J. F. (2015). *Migration Theory. Talking Across Disciplines* (3rd ed.). New York: Routledge.

Caglar, A. & Glick Schiller, N. (2018). *Migrants & City-Making. Dispossession, Displacement & Urban Regeneration*. Durham, NC: Duke University Press.

Castles, S. & Miller, M. J. (2009). *The Age of Migration. International Population Movements in the Modern World*. Basingstoke, UK: Palgrave Macmillan.

Christ, O. et al. (2014). Contextual Effect of Positive Intergroup Contact on Outgroup Prejudice. *PNAS*, 111(11), 3996–4000.

Clark, C. D. (2011). Doing Child-Centered Qualitative Research. In Clark, C. D. (Ed.) *A Younger Voice* (pp. 1–130). Oxford: Oxford University Press.

Crate, S. A. (2011). Climate and Culture: Anthropology in the Era of Contemporary Climate Change. *Annual Review of Anthropology*, 40, 175–194.

Delgado, M. (2006). *Designs and Methods for Youth-Led Research*. Thousand Oaks, CA: Sage.

Dimitrova, R. (2017). *Well-Being of Youth and Emerging Adults Across Cultures. Novel Approaches and Findings from Europe, Asia, Africa and America*. Cham: Springer.

Dimova- Cookson, M. & Stirk, P. M. R. (2010). *Multiculturalism and Moral Conflict*. London: Routledge.

Emerson, R. M., Fretz, R., & Shaw, L. L. (1995). *Writing Ethnographic Fieldnotes*. Chicago, IL: University of Chicago Press.

Ensor, M. O. & Gozdziak, E. M. (2010). *Children and Migration. At the Crossroads of Resiliency and Vulnerability*. Basingstoke, UK: Palgrave Macmillan.

Eriksen, T. H. (2009). Living in an Overheated World: Otherness as a Universal Condition. In: Identity Politics: Histories, Regions and Borderlands. *Studia Anthropologica*, III, 9–24.

Eriksen, T. H. (2010 [1995]). *Small Places, Large Issues. An Introduction to Social and Cultural Anthropology*. London: Pluto Press.

Eriksen, T. H. (2017). Global Citizenship and the Challenge from Cultural Relativism. *Issues in Ethnology and Anthropology*, 12(4), 1141–1151.

European Commission. (2017). Youth Policies in Finland 2017. *Youth Wiki National Description*. Retrieved from: https://eacea.ec.europa.eu/national-policies/sites/youthwiki/files/gdlfinland.pdf

Finnish Government Report. (2018). Finland. Executive Summary of the Government's Report on the Implementation of the 2030 Agenda for Sustainable Development. Retrieved from: https://kestavakehitys.fi/documents/2167391/2186383/Executive_Summary_2018_v1.2.pdf/9bdb30cc-bfa6-4692-8922-424a2fb0dd27/Executive_Summary_2018_v1.2.pdf.pdf

Flick, U. (2014 [1998]). *An Introduction to Qualitative Research* (5th ed.). London: Sage.

Fortier, A. M. (2000). *Migrant Belongings. Memory, Space, Identity.* Oxford: Berg.

Fraser, S., Lewis, V., Ding, S., Kellett, M., & Robinson, C. (2004). *Doing Research with Children and Young People.* London: Sage.

Hammar, T., Brochmann, G., Tamas, K., & Faist, T. (1997). *International Migration, Immobility and Development: Multidisciplinary Perspectives.* Oxford: Berg.

Heikkilä, E. & Peltonen, S. (2002). *Immigrants and Integration in Finland. Report.* Turku, Finland: Institute of Migration. Retrieved from: http://maine.utu.fi/articles/069_Heikkila-Peltonen.pdf

Heikkilä, M. (2019). *If We Lose the Arctic. Finland's Arctic Thinking from the 1980s to Present Day.* Rovaniemi: The Arctic Centre, University of Lapland.

Heikkinen, M. P. (2019). Haavemaa [Dreamland]. In: Napapiirin Tulevaisuus [The Future of the Polar Circle]. *Helsingin Sanomien Teema-lehti, 1,* 22–37.

Heleniak, T. (2018). Migration. The Wary Welcome of New-comers to the Nordic Region. In Grunfelder, J., L. Rispling, & G. Norlén (Eds.) *State of the Nordic Region 2018. Nordic Council of Ministers.* (pp. 48–58). Denmark: Rosendahls.

Hirvi, L. (2013). *Identities in Practice. A Trans-Atlantic Ethnography of Sikh Immigrants in Finland and in California.* Helsinki: SKS.

Jokela, T. & Coutts, G. (2017). *Relate North. Culture, Community and Communication.* Rovaniemi: Lapland University Press.

Kehily, M. J. (2007). *Understanding Youth: Perspective, Identities and Practice.* Los Angeles, CA: Sage.

Kokot, W. & Dracklé, D. (1996). *Ethnologie Europas: Grenzen, Konflikte, Identitäten* [Ethnology of Europe: Boarders, Conflicts, Identities]. Berlin: Reimer Verlag.

Koser, K. (2016). *International Migration: A Very Short Introduction.* Oxford: Oxford University Press.

Kuhmonen, T., Kuhmonen, I., & Luoto, L. (2016). How Do Rural Areas Profile in the Futures Dreams by the Finnish Youth? *Journal of Rural Studies, 44,* 89–100.

Lancy, D. F. (2008). *The Anthropology of Childhood. Cherubs, Chattel, Changelings.* Cambridge: Cambridge University Press.

Larsen, J. N. & Fondahl, G. (2014). *Second Arctic Human Development Report.* Regional Processes and Global Linkages. Retrieved from: http://norden.diva-portal.org/smash/get/diva2:788965/FULLTEXT03.pdf

Larsen, J. N., Schweitzer, P., & Petrov, A. (2014). *Arctic Social Indicators II.* Nordic Council of Ministers. Retrieved from: www.sdwg.org/wp-content/uploads/2016/04/Arctic-Social-Indicators-II.pdf

Määttä, K. & Uusiautti, S. (2019). Arctic Education in the Future. In Uusiautti, S. & N. Yeasmin (Eds.) *Human Migration in the Arctic. The Past, Present, and Future* (pp. 213–238). Singapore: Palgrave Macmillan.

Martin, P. (2015). Economic Aspects of Migration. In Brettell, C. B. & J. F. Hollifield (Eds.) *Migration Theory. Talking across Disciplines* (3rd ed.), pp. 90–114). New York: Routledge.

Mathews, G. & Izquierdo, C. (2009). *Pursuit of Happiness. Well-Being Anthropological Perspective.* New York: Berghahn Books.

Matthews, H., Limb, M., & Taylor, M. (1999). Young People's Participation and Representation in Society. *Geoforum, 30,* 135–144.

McAuliffe, M. & Ruhs, M. (2017). *World Migration Report 2018.* Geneva: International Organization for Migration. Retrieved from: www.iom.int/sites/default/files/country/docs/china/r5_world_migration_report_2018_en.pdf

Meissner, F. & Vertovec, S. (2015). Comparing Super Diversity. *Ethnic and Racial Studies, 38*(4), 541–555.

Moore, J. D. (2009). *Visions of Culture. An Annotated Reader.* Lanham, MD: AltaMira Press.

Olwig, K. F. & Hastrup, K. (1997). *Siting Culture. The Shifting Anthropological Object*. London: Routledge.

Petäjämaa, M. (2013). Lapin Maahanmuuttostrategia 2017 [Strategy for Finnish Lapland Immigration 2017]. *Lapin elinkeino-, liikenne- ja ympäristökeskus*. Retrieved from: www.doria.fi/bitstream/handle/10024/87882/Elinvoimaa-alueelle-1-2013.pdf

Petersen, H. & Poppel, B. (1999). *Dependency, Autonomy, Sustainability in the Arctic*. Aldershot, UK: Ashgate.

Petrov, A. N., Burn, S., Shauna, C., Stuart, F., Fondahl, G., Graybill, J., Keil, K., Nilsson, A. E., Riedlsperger, R., & Schweitzer, P. (2017). *Arctic Sustainability Research. Past, Present and Future*. London: Routledge.

Pollock, G., Ozan, J., Goswami, H., Rees, G., & Stasulane, A. (2018). *Measuring Youth Well-being. How a Pan-European Longitudinal Survey Can Improve Policy*. Cham: Springer.

Silverman, D. (2013). *Doing Qualitative Research* (4th ed.). London: Sage.

Silverman, D. (2014). *Interpreting Qualitative Data* (5th ed.). London: Sage.

Stammler, F. & Adams, R. M. (2019). Arctic Centre Project: Live, Work or Leave? Youth – Wellbeing and the Viability of (Post) Extractive Arctic Industrial Cities in Finland and Russia. Unpublished raw data.

State Youth Council. (2019). Youth Barometers. Retrieved from: https://tietoanuorista.fi/en/publications/

Suárez-Orozco, M. M. et al. (2011). Migrations and Schooling. *Annual Review Anthropology*, 40, 311–328.

Taskiainen, S., Uusiautti, S., & Määttä, K. (2019). How to Enhance Immigrant Students' Participation in Arctic Schools? In Uusiautti, S. & N. Yeasmin (Eds.) *Human Migration in the Arctic. The Past, Present, and Future* (pp. 143–169). Singapore: Palgrave Macmillan.

TEM (Työ- ja elinkeinoministeriö). (2019). *Maahanmuuttajat Lapissa* [Immigrants in Lapland]. Retrieved from: https://kotouttaminen.fi/maahanmuuttajat-lapissa

Tsuda, T., Tapias, M. & Escandell, X. (2014). Locating the Global in Transnational Ethnography. *Journal of Contemporary Ethnography*, 43(2), 123–147.

Uusiautti, S. & Yeasmin, N. (2019). *Human Migration in the Arctic. The Past, Present, and Future*. Singapore: Palgrave Macmillan.

Vertovec, S. (2001). Transnationalism and Identity. *Journal of Ethnic and Migration Studies*, 27(4), 573–582.

Vertovec, S. (2009). *Transnationalism*. London: Routledge.

Vertovec, S. (2010). Towards Post-multiculturalism? Changing Communities, Conditions and Contexts of Diversity. *International Social Science Journal*, 61, 83–95.

Vertovec, S. (2011). The Cultural Politics of Nation and Migration. *The Annual Review of Anthropology*, 40, 241–256.

Vertovec, S. (2012). "Diversity" and the Social Imaginary. *Archives European Sociology*, LIII(3), 287–312.

Vertovec, S. (2017). Mooring, Migration Milieus and Complex Explanations. *Ethnic and Racial Studies*, 40(9), 1574–1581.

Vertovec, S. (2019). Talking Around Super-diversity. *Ethnic and Racial Studies*, 42(1), 125–139.

Yeasmin, N. (2012). Life as an Immigrant in Rovaniemi, Finland. In Tennberg, M. (Ed.) *Politics of Development in the Barents Region* (pp. 340–361). Rovaniemi: Lapland University Press.

Part III

Family and diversity challenges

5 Migrant integration in Finland

Learning processes of immigrant women

Nafisa Yeasmin and Stefan Kirchner

Introduction

Compared with the countries of Central and Southern Europe, Finland has seen relatively modest levels of immigration, both in terms of absolute numbers of immigrants and with regard to the percentage of immigrants as a portion of the overall population.

The integration of migrants into society can be particularly challenging for immigrant women who might face different barriers to integration. There are many many hurdles for integration that immigrant women face. However, Finnish language is a one of the major impediments that hinders their fast integration process. To ensure social inclusion of immigrant women, good policies need to be in assessed. More support to empower this group and aid their learning process can create opportunities for them to feel potential citizens in the host country.

Many of them lack a formal education from their country of origin, which limits their access to the local labour market, and language skills limit their access to local formal and informal networks accordingly. Many times, cultural differences between the country of origin and the host country hinder their integration process.

In this study, we initiated exploration of their whole integration path, which is basically a learning process for them. Assessment of learning includes information about the group who are being assessed. It is necessary to have an in-depth understanding of the group members or individual members of the group. The group of learners need a constructive support method based on their learning needs. The study explored whether their integration indicates a transformative learning process, whether they are motivated enough to learn more for their resettlement in Finnish Lapland. Application of self-determination theory gives a deeper understanding of the difficulties they face balancing their gender dynamic within the intercultural context.

In this chapter, we will look not only at theoretical background questions regarding the integration of immigrant women in Finland, but also at the practical challenges faced by women who are migrating to the country. To this end, the empirical aspects of this chapter are based on six interviews with immigrant women and the views of an ethnographic observation. The chapter aims to answer the question of how the integration of immigrant women can be improved, and how learning can empower immigrant women to play a more active role in their integration into Finnish society.

Methodology

Methods

The research was based on qualitative research, using small, purposeful semi-structured interviews, the experiences of six interviewees, and the authors' ethnographic observations, working with this group of people from 2018 to 2019. Interviews lasted for a maximum of 2 hours each. These face-to-face interviews were transcribed on paper. All interviewees had a refugee background and were from Asian and Mediterranean countries. The interviews were organized around three thematic axes, one of which concerned overall integration and learning about integration. Their feelings, experiences, changing situation, social participation, keeping of customs and traditions, assimilation process, and friendship circles were all observed for comprehensibility across research validity and coded in order to make the research results reliable. Those codes are the most significant and relevant data we explored to conduct our further analysis. Our arguments are illustrated based on observational data as well.

This study is part of an ethnographic observation (EO) in which one of the authors has analyzed unstructured data that were collected directly from the field and described on the basis of observation of the cultural situation of the research group. The author had the opportunity to participate directly in different conversations and family occasions with various immigrant women. Materials were collected by attending regular community events. Basically, observations of the women's culture, customs, and traditional lifestyles that are illustrated were based on the researcher's experiences at the time of those conversations and during participation in events. It is also a narrative-based interpretation of such events and meetings (Dey, 2002).

However, the limitation of this research was the need to ensure ethical treatment of our informants, whether we were experienced enough to illustrate understanding of the views of the group properly, as, ethically, researchers have to avoid subjectivity and positionality in many areas, which also constrains them to explain many issues that are considered a discomfort zone for participants. There is constant pressure for researchers to avoid ethical and culturally sensitive issues in such EO (Huddle, 2018). The reliability and validity of such research are under potential threat if researchers lack the ability to avoid the natural settings of purposive manipulation. However, the results of this research are adapted within a broader narrative and descriptive design (LeCompte, 1982), which are supported not only by research findings, but also by literature review. Such threats are avoided by interviewing immigrant women in the study and codifying the techniques of reliability by choosing thematic axes mentioned before and selecting subthemes for investigating research validity. The research reliability was also supported by clarification of previous relevant research.

Theoretical context of integration of immigrant women

Finland's integration programme is modern and flexible; it includes different integration measures and resources for successful integration (GIP, 2016–2019). The government programme focuses on four thematic axes: (1) using immigrants cultural strengths to ease integration; (2) encompassing cross-sectoral measures to enhance integration; (3) more cooperation between state and municipality; and (4) more discussion on immigration policy against racism (GIP, 2016–2019). The second thematic axis includes improving access to the labour market. The employment rate of native-born women is 68.8 per cent, which is higher than the OECD average of 67.4 per cent – the average among the member states of the Organisation for Economic Co-operation and Development (OECD), a group of 36 developed countries, of which Finland is a member. If the participation rate is higher, then the unemployment rate is lower. But the participation rate is comparatively lower among immigrant foreign-born women in Finland (OECD, 2018). Basically, the diversity of the different cultural, geographic, religious, ethnic, professional, and other backgrounds of immigrant women also has an impact on integration. Accordingly, the acculturation processes of women with refugee status differ from those of women who arrived as labour migrants or as a spouse of a Finn, although most evidence of integration action plans and strategies is similar for all groups of women. But some of the groups, especially refugee women, are very much dependent on settled integration measures rather than engaging in self-integration. On the other hand, very little attention has been paid to integration measures and strategies with a particular focus on immigrant women. Many different actions and strategies are needed, some of which could be focused on the gender issues of immigrant women (EU, 2018).

As Finland is facing challenges in integrating immigrant women, a women-specific (EU, 2018) integration framework can be targeted for specific groups who have specific needs and depend on settled integration measures. Many countries have adopted a women-specific policy, but harmonization of this integration module in practice is rather challenging, as the groups are heterogeneous, and their needs are different – for example, some are hard-to-reach immigrant women. So, identifying problems related to the settlement of immigrant women needs more attention in the implementation of policy. Here, "settled integration" refers to the local and national settlement policies and measures, which differ very much between Finnish municipalities, because they have, to a great extent, autonomous characteristics. Some of the national policies do not create any legal obligations for municipalities. Therefore, local settlement policies differ between municipalities, which has also an impact on overall integration. Many cities provide many more integration activities for immigrants based on their individual municipality resources. As Finland has adopted a multiculturalist policy that is very much based on Nordic evidence (Saukkonen, 2016), settled integration is based on service needs along with an initial assessment of an individual who has already received a residence permit. Immediately after the assessment, a mutual integration plan starts (ibid.) that aims at improving the immigrant's different skills that are deemed necessary for their successful integration into Finnish society. In reality, the practical implementation of some integration activities that are included in the integration plan or programme does not always take place immediately, in all municipalities, at the same time (ibid.). On the other hand, some of the services that are provided exclude gender issues to some extent.

According to our research, our target group lacks an expected "settled integration" model that can oblige them to integrate into Finnish society. As a result, it turns into a self-integration model that is flexible, based on individual characteristics, family, and cultural opportunities. Such a model of self-integration creates some signs of ethnic segregation and self-isolation, which can hinder the integration process in several respects.

The self-integration model very much depends on self-determination – whether a person is motivated to learn more about life management skills in a new social environment. Immigrant women can increase their motivation though boosting competencies for managing their new life, which needs powerful, individual control of their own behaviour and need to create a sense of belongingness to connect with people in the host society.

Our acculturation research on the motivation of immigrant women to learn in a new society shows a challenge-based integration that depends on both settled and self-integration. However, we argue that both types of integration process need an initial motivation for acculturation behaviours.

In a new environment, immigrant women need a supportive, constant learning process that can facilitate knowledge transformation. Self-determination theory (SDT) gives an understanding of how to operationalize a motivational strategy within a group by setting different implementations. This has been applied to motivate children in education (Kathryn et al., 2016; Ryan & Deci, 2016; Deci et al., 1996; Deci & Ryan, 2002; Parker & Seal, 1996), SDT of teachers and parents (Legault et al., 2006; Pelletier & Sharp, 2009; Alivernini & Lucidi, 2011; Bowers & Sprott, 2012; Fall & Roberts, 2012), and the health behaviour of different groups of people (Wilson et al., 2006; Webb et al., 2010; Dombrowski et al., 2014; Moore et al., 2015; Gagnon et al., 2018) for the purpose of increasing their level of self-determination. Our research also targets the establishment of a connection with SDT in the integration of vulnerable immigrant women in the host society. As immigrant women are often poorly integrated (OECD, 2018), there is a need to increase the capacity to support autonomy, attaining competencies for developing participation skills in the host society and learning appropriate methods for identifying opportunity structures to ease their acculturation process. This is quite a long learning process to control one's behaviour (Fortier et al., 2012; Mata et al., 2009, Su & Reeve, 2011; Cesaroli et al., 2014, 2012), leading to evocative activities facilitating immigrant women's self-endorsement (Teixeira et al., 2012) and feelings of the sense of belonging to the host society. They need support for attaining knowledge concerning the social structure of the host country, instead of depending on partners or other family members along with the extended family (Yeasmin & Koivurova, 2019). There are many settled integration measures by which to identify which is the better option for immigrant women to regulate contingencies of interest, engagement, and motivation.

Indeed, women alone cannot regulate their self-determination boosting at the earlier stage: they need external support from the social ecological system of the host country:

> women's socio-cultural adaptation is slower than immigrant men's due to both acculturative stress and individual stressors. Immigrant women are minorities among a minority, and the integration of minorities in such a sparsely-populated territory demands national identity and a sense of belonging in a territory.
>
> (Yeasmin & Koivurova, 2019)

For this, they need motivation to feel connectedness with the majority society.

Findings

Contextual analysis of the study (see Figure 5.1)

Autonomy

A women-specific integration model encompasses an autonomy-supportive concept. Immigrant women need support to control the courses of their new life in the host society. They need to develop their psychological needs to strengthen their motivation for improving their cohesive sense of self (Deci & Ryan, 1985, 2000; Kendra, 2017). According to the analysis of our data, immigrant women

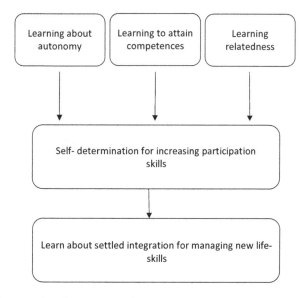

Figure 5.1 Contextual analysis process of SDT

(e.g. refugees from different Muslim countries) are dependent on the construction of their families. Their family configurations reproduce and reinforce ethnocultural segregation (Block, 2015). Family priority is potentially more important than personal autonomy, which can lead to individuals becoming very much focused on family traditions rather than on being integrated into a new society (Haan, 2008). The focus on family tradition can mean that immigrants prioritize families and friends (ibid.) over integration. Many immigrant women even prioritize the traditions and customs of transnational extended family whom they have left behind.

"Our common ethnic origin is strong factor that unified us in many extents, though our nuclear family lives in this city, but our extended family that we left behind has a strong impact in our everyday life" (F, 35–55).

Many nuclear family decisions depend on common values of transnational extended family (Telegdi-Csetri & Ducu, 2016). A strong, continuous relationship with extended family members reflects on the construction of nuclear family life in the host society (EO, 2018–2019), although they live in different nation-states. Many nuclear family challenges linked to gender dynamics and cultural and religious conflicts in immigrant families are usually shared with transnational extended family for mutual solutions (EO, 2018–2019).

Opportunities for immigrant women to control the mutual adaptation process vary from family to family. In many families, immigrant women can control the process of building adaptive capacities with others, which facilitates autonomy support for individual development in a new environment (Nelson et al., 2007; EO, 2018–2019). If immigrant women receive support from their family in order to educate themselves about the social and cultural life of the host country, it can enhance their probable determinants to assimilate in a new environment (EO, 2018–2019). All immigrants face particular challenges to get access to every sphere of the host society. Their life-management skills are contingent on their experiences of and education in the social life of the host country. These skills also ease the difficulty of contending with multiple forms of racism. Following family tradition, on the other hand, can lead to alienation rather than enabling integration (Wray, 2009; Social Planning Council of Ottawa, 2010; Spitzer et al., 2012; Craig, 2015; EO, 2018–2019). Immigrant women from various ethnic backgrounds ascertain factors and obstacles to their autonomization in different phases of their integration (Daniélé, 2016; EO, 2018–2019).

Immigrant women find themselves between their more family-oriented traditions and the traditions of the host countries and have to navigate this area of tension in their attempt to achieve successful integration (EO, 2018–2019). In many cases, this conflict can hinder or reduce their capacity to make decisions to increase their participation level in host societal activities (Allan, 1998). In different cultural-conflict situations, when they are in identity crisis, they are confused about representing their power dynamics of autonomy, as they cannot make decisions on such a new circumstance related to integration without discussing it with either their nuclear or transnational extended family (EO, 2018–2019).

Immigrant women influenced by the culture of their country of origin create a situational logic to act impassively to a certain extent (Daniélé, 2016). They may lack the self-reflexivity (Archer, 2007) that is required to underpin the meaning of integration and the development of autonomy in the face of different obstacles. According to previous studies, they lack autonomy because of their low educational status, cultural image, interpretation of gender issues, and position in the society of origin (Ebrahim, 2017). In many immigrant families, men are in the position of power, which contributes to a lack self-reflexivity in women (Ebrahim, 2017) to enact autonomy. It sometimes constrains women's ability to enjoy their freedom and sovereignty in the acculturation process (Daniélé, 2016; Roxanne, 2010). Some immigrant women's social participation and relations with the surrounding society are affected by their transnational family experiences (EO, 2018–2019). Communities of male pioneers who are living in different nation-states to their extended family, their sentiments, and family roles and responsibilities strongly associated with previous experience. However, very little changes have been observed with second-generation immigrant women (EO, 2018–2019).

Competence and competencies

Integration patterns as such are discursive (Saukkonen, 2016), which, to a certain extent, demands a set of self-competencies in a person who wants to be integrated into the host society. Local language skills are necessary to communicate with others in Finland or any other of the Nordic countries. Training programmes provides a package of materials to familiarize immigrants with integration policies, including information about the labour market, communities, culture, and the sociopolitical life of the host society. Therefore, immigrants need competencies that are to be linked with activities provided for the purpose of achieving proper integration. The competencies usually build on the immigrant's individual personability and motivation (James, 1890). For immigrant women to gain competencies depends on the nature of competence-motivated behaviour. The competencies that immigrants have with regard to the traditional cultural values, views, and behaviour of their country of origin are not necessarily applicable and desirable when communicating with others (Sue, 1999, 2003; Bernal & Scharrón-del Río, 2001; Hall, 2003). They need to acquire knowledge from a different dimension likely to increase the level of acculturation, which necessitates interaction with other agents in new societies, sharing gender issues and experiences between them. They need to build their competencies. The previous experiences and personal know-how brought by immigrant women from their country of origin are indeed a form of human capital that immigrant women need to manage within a strategic framework in order to evaluate their determinants for their own integration into a new environment.

Individual competencies are comprised of a self-motivated creativeness to acquire knowledge, skills, and abilities (Klein, 1998; Beardwell et al., 2001; Martin, 2010). Immigrants need to give a new direction to their competencies within a new and unfamiliar situation, which can change the course of life by replacing certain values for the purpose of boosting one's self-esteem. Competencies that can build the self-determination of immigrant women and enable equal capacities and expectations are part of a changing adaptation framework. Integration is a process of renewing and adjusting competencies (Johnson, 1981). Proper integration is required not only in order to fulfil societal demands, but also for better integration of the children of immigrants (OECD, 2018). The competence to know "special issues are associated with health and well-being is an important part of integration" (THL, 2018, n.p). There are many factors that have effects on health, such as untreated illness and lack of access to treatment services, a diet according to the environment, health care, and information about quality of life. All of these issues demand competency to find all those available services and language skills (THL, 2018). In this regard, refugees and their families are in a vulnerable position. In many cases, lack of knowledge on the mother's part has a negative impact on children's healthcare and school integration (OECD, 2018). Healthcare systems also vary from country to country in terms of genderplay treatment. In many cases, the cultural influence of the country of origin hinders immigrants' perceptions about help-seeking behaviour, which also has an impact on coping style (O'Mahony & Donnelly, 2007; Guruge et al., 2008; Adler et al., 2010; Donnelly et al., 2011; EO, 2018–2019). Immigrant women are dependent on male privilege to understand the bureaucratic process and functioning of services. Religious and cultural obstacles (Douki et al., 2007; EO, 2018–2019) can prevent immigrant women from sharing their health stories with professional health workers, and the idea of bringing shame on the family also creates barriers to receiving healthcare. Women need to learn about all the facilities, availabilities, and accessibilities to overcome all these structural barriers that hinder their adaptation. Immigrant women who are victims of family violence (Lee & Hadeed, 2009) need to learn about the laws of the host country, which demands competencies to develop certain kinds of experience (EO, 2018–2019).

Relatedness

Relatedness is a basic human need. Feelings of relatedness or being connected with others can boost the motivation of individuals. How attached immigrant women are to the host society has long-term effects. In the case of children's motivational development, relatedness to peers and healthy relationships matter in school performances (Bretherton, 1985; Crittenden, 1990). Sensitive and responsive interaction can secure attachments with peers. Social contact and trusted support allow people to be more engaged in social activities and increase interaction with others.

A sense of relatedness is a factor that is related to the social integration of all immigrants. Integration very much depends on how easily immigrants and their family can be incorporated into the host society. Feeling "at home" is a psychological identity that is tied to a particular place (Sigmon et al., 2002). A sense of belonging to a particular place is very much spiritual and emotional (Duncan & Lambert, 2004), interconnected with private and public memories, sociocultural norms, and home traditions. According to earlier research, "[t]he immigrant is expected to gradually release previous attachments, social identifiers and even a sense of national commitment to his country of origin, and develop a sense of local identity and belonging in the host country" (Amit & Bar-Lev, 2014: 948; EO, 2018–2019).

It is comparatively easier for immigrant men to cope with the tension between new identities and old cultural affinities with their country of origin. There are certain push factors that support immigrant men in constructing their new identity (Lerner et al., 2007) As their patriarchal family background pushes them to adapt, this helps immigrant men in their autonomization process and in achieving competencies (EO, 2018–2019). As immigrant men from patriarchal societies are often occupied with gaining access to the labour market, which requires collecting information for the purpose of improving their quality of life, they are often committed to stay in the host country. This enables immigrant men to sustain membership in multiple communities (Levitt, 2003; EO, 2018–2019). They are seen as significantly optimistic and as responsible for improving their participation in the host country.

On the other hand, the relatedness of immigrant women to the host society is comparatively lower than in the case of men (Yeasmin & Koivurova, 2019). Immigrant women often have a strong sense of belongingness to their source country, which can slow down the integration process. Source country attributes are important among immigrant women, which can hinder their interactions with the host society. Immigrant women hypothesize a natural link with their country of origin (Ghorashi, 2017; Yeasmin & Koivurova, 2019). On the other hand, a lack of autonomy and competencies can lead to self-isolation.

The practice of extending regular interaction with the members of families who live in the destination country raises a natural psychological link, and members try to keep in touch with the family back in the home country or family who are currently in the process of moving from one country to another through digital support to create, exchange and reiterate their sense of belongingness to the family, as it also give them emotional support when they can share similar values (Kilkey & Palenga-Möllenbeck, 2016; EO, 2018–2019).

The impact maintaining transnational family relationships has on establishing a traditional gendered approach in acculturation processes is visible in much previous research (Schneebaum et al., 2015; Fresnoza-Flot & Shinozaki, 2017; Marchetti & Salih, 2017; Ala-Mantila & Fleischmann, 2017). It is much easier for migrants to follow a strategy of living with self-created community by accepting a self-integration method rather than not to display themselves in the social category discussion, not living under the pressure of integration, not fighting discrimination (Ducu, 2018).

> When we came to this city as newcomers, we tried to find people who shared the same customs and tradition as well as language. Social connections with such people who shared similar ethnic background gives feelings of safety network. … There are no such families who speaks same languages then tried to contract with Finnish neighbors – they are friendly, but don't feel a kinship tie with them. I cannot express all my feelings with Finnish neighbours but I can share everything with my transitional family who support me in every steps.
>
> (Female immigrant)

Many such cultural and personal attributes (Yeasmin & Koivurova, 2019) hinder the willingness of immigrant women to integrate into the host society and do not create any obligation or relatedness to the host society.

Challenges of learning

Some potential challenges that were identified in this research are (1) cultural loneliness of immigrant women in Finland, (2) inability to communicate in the local language, and (3) dominant enculturation. According to the 2018 OECD report, immigrant women in Finland are a relatively poorly integrated group: immigrant women are seen as not utilizing social activities and mostly remaining outside the labour market, which can have negative impacts regarding the long-term integration of their children into Finnish society (OECD, 2018). However, the integration programme for the target group is flexible. Acculturation could influence the well-being of immigrant women in the North. The acculturation process is influenced by many factors. Some factors can be controlled by individual behaviour, and some are controlled by socio-environmental behaviour (Bandura, 1977). Muslim immigrants can face some challenges when pre-immigration and post-immigration experiences differ enormously from each other, both theoretically and empirically. This transition can hinder the acculturation process towards a new socio-ecological perspective. The sociocultural factors of two different societies have an impact on acculturation in a new environment. Social connections, actions, and interactions have a powerful effect on acculturation as well as social learning systems. Social connections have an effect on social learning systems, and, relatively, social connections and social learning systems have an effect on acculturation.

> Life situation is different, life was social enough in my country of origin. Friends, relative are always around us; Frequently we meet each other's – almost everyday relationship with them as well as neighbors. There were many activities among women – this is missing here.
>
> (F, 30–50)

As well as the social network, network support can also ease acculturation.

It does not mean that immigrant women need network support from their own community, but they need support from the host society. Such support can enhance the immigrant's social learning capability and motivation.

> We speak our own languages when we discuss with our peers, that doesn't help to learn local language and such discussion doesn't get that much information about local communities, but give cognitive appraisal of being connected with someone at least. One of my good friends already moved from this city.
>
> (F, 35–55)

Social ties or bonds are one reason for successful integration. If the social ties are poor, this causes poor integration into Finnish society (Forsander, 2008).

The relationship of immigrant women with their neighbourhood or any kind of voluntary organization can also support social integration; the absence of some activities also makes immigrant women vulnerable and can hinder their integration. Muslim women are under-represented in the labour market (Chang & Holm, 2017), rarely hold management positions, lack social roles, and often earn less than their male counterparts (Sarvimäki, 2011). The overall employment opportunities for immigrant women, in particular in the labour market, are even worse (Sarvimäki & Hämäläinen, 2010). The stress of integration and gender discrimination among families can also hinder the balancing of self-esteem and sense of identity.

> Lack of social support or social influence just create a loneliness and feelings of social exclusion in me. Do not feel any community obligation any more. Less opportunities for sociability and cultural activities make us socially vulnerable group. Stress of integration, differences with respect to gender role between this two societies has an influential effect on my life ... since, gender perspectives are different in my family and host societies.
>
> (F, 30–50)

Ethnic identities and lack of language skills limit social interaction and the ability to communicate. Some of the families follow patriarchal patterns of lives, which also hinders women's empowerment in the host society and causes poor integration results. To a great extent, they are dependent on their family for all decision-making. This creates self-exclusion, which can affect their integration. Lack of family influences causes a lack of determination in immigrant women to be integrated into a new culture.

> I didn't work in my home country and I have many family responsibilities that doesn't support me to work here. I have to follow my traditional norms and culture which is acceptable by our own community, I cannot do such activities that is not accepted by our ethnic community members.
>
> (F, 30–50)

Immigrants' ethnic communities are so small in the North when compared with Southern or Central Europe. Therefore, it is important for many immigrants to adhere to their own traditional norms and cultures in order to maintain the limited ties between small communities. As the limited contacts outside their communities create a fear of community exclusion among them, this can encourage them to abstain from adopting the cultural norms of host societies in order to maintain a relationship with their own community and peers.

Discussion and conclusion

So far, SDT has been considered more applicable to children's learning in school. We have chosen this theoretical approach here to look at the learning of immigrant women about their new future in the host country. SDT supports the sense of learning within a time frame through building a successive method of sharing experiences, creating autonomy, and developing a sense of willingness, which usually develop children's learning opportunities as a whole. Here, we also believe that SDT could equally be applied in order to increase immigrant women's capability during the integration process to make them aware of their integration.

A women-specific integration strategy can include activities that are related to their values and create a sense of willingness. Such activities that are enjoyable for immigrant women can create self-learning capacities and also improve their motivation owing to associations with positive learning outcomes. Immigrant women may need instruction with regard to self-management skills, which needs to include various training related to daily, social living in a new environment.

> Usually, when immigrant women are invited to meet social worker for their regular follow up meeting, many women go with husband. In this case, husband talks on behalf of the wife with social worker. Such kind of provision should not be acceptable. It does not support women's autonomy and not support indeed to be decisive.
>
> (EO, 2018)

Many immigrant women are not given the opportunity to learn problem-solving skills and measuring skills on self-report assessment. Strategies inspired by the needs of women should allow them to live more independently. Family norms and customs and traditions that they practise at home are not autonomy-supportive, which creates obstacles to self-selected goals. A goal-attainment social policy that involves women in order to help them establish their own goals regarding their future competencies and relatedness can specify a range of outcomes in the host country.

> [M]any immigrant women chose their vocational educational place based on their family decision and in many families, women are not permitted to work in a working place where there are men as a worker – which hinder their scope of job opportunities.
>
> (EO, 2018)

To some extent, immigrant women cannot actively participate in social activities, as it is the tradition in various families that the childcare is the responsibility of women alone. Many women even lack specific information that they could use in preparation of their well-being: for example, in Nordic countries such as Finland, men can also take paternity leave and fulfil childcare duties. Such information that can support women's integration can be distributed among immigrants in specific ways so that they can easily access the information – for example, through regular information targeting immigrant women. Adjusting actions that can set the goal to meet the needs of immigrant women to construct their learning environment is significantly important and should be included in women-specific integration strategies.

Many women also lack gender-specific knowledge, which limits their ability to be the causal agent of their own lives. Taking care of their children usually excludes them from integration training after arriving in a host country (OECD, 2018). In many welfare countries, mothers receive child care benefits, which influences families when deciding on the allocation of household activities, often with a preference for women staying at home in traditional family configurations. Therefore, many immigrant women cannot complete their integration training, which includes learning many skills, such as local language skills, sociocultural knowledge, knowledge of fundamental national laws and practices, along with obligations and responsibilities.

"I did not even know earlier that men are also responsible for childcare. In our country, usually women take care of children at home and men do job. I don't have any job experiences from my country of origin" (F, 35–55).

Hence, an integration strategy directed at women could provide immigrant women with new perspectives that can enable them to gain self-competence and can make them determined to be integrated into the host society. Social participation support can give them the feeling that they are related to a system and an integral part of the system relatedness.

Different policies that are aiming to enhance the sense of belongingness of the target group necessitate acquiring specific knowledge for both formal and informal service providers so that they can navigate women towards specific settlement services. It can be necessary to offer more learning opportunities to know about their neighbourhood and recreational facilities.

A lot of activities are offered to immigrant women that do not create interest among the target groups. Therefore, immigrant women's involvement in such activities is often relatively low. The programmes and activities can be designed by immigrant women. This can create a platform for immigrant women to plan and design programmes that are suitable for them. At the same time, immigrant women need to be given a place where they can learn, for better acculturation. Not only participation in activities but also involvement in the planning of social activities can give immigrants a feeling of belonging to a group. They can learn more when they feel that they are not only the target group, but they are also useful for the target group. This also creates self-confidence and self-competencies and it could also be a cross-cultural learning platform with sharing individual experiences and involvement. If a society were to recruit immigrant women as volunteers or through job contacts in leading roles, it would be an example of autonomy-supportive measures that also facilitate increased autonomy development that allows them to maintain their own family relationships.

Though previous research has shown that the integration of migrants is a lifelong learning process (Guo, 2013), it can be mediated through social interaction and by sharing the context of the lived experiences. Each of these experiences within social networks and relationships provides a context of understanding learning. A multicultural and multidimensional network can provide immigrant women with learning opportunities. Institutional learning or integration training could be practical for immigrant women if it includes more learning methods that allow them to increase self-determination, as well as learning that takes place in the community (Alfred, 2013).

The 2018 OECD report also states that a lack of ability to communicate with school staff hinders immigrant women's participation in their children's educational experiences. Learning the local language is an important part of integration as it facilitates interactions with the host country's society as a whole. Creating self-motivation by combining language learning with a proper (as seen from the host society's perspective) understanding of better family life might ensure women's successful participation in the language learning process. A psychological gender-based empowerment tool for adaptation to and familiarization with the host society is a key aspect for many of the interviewees.

> I used to go to day care Centre with my child and tried to observe how my child is learning language and taking part in group activities, which also helped me to learn language. However, teachers in the day care Centre didn't like my such approach to be with child in the day care for long. However, I enjoyed that situation of learning.
>
> (F, 30–50)

It follows that all kinds of learning methods that facilitate transformation of experiences (e.g. by feeling, watching) towards a new experience are argued to be a proper learning method by Kolb and Fry (1974). The main objective of social learning is to create an interest in learning among immigrant women. There are no logical approaches for social learning, which is an emotional influence, such as paying attention, observing situations, and learning through conditioning (Bandura, 1977). It is less relevant how a person is learning; rather, it is the determination to learn that plays an important role, because motivation and self-determination may be enough for reinforcement in learning where the target group is marginalized and at risk of various difficulties during the process of acculturation. Social learning theories sometimes suggest that role models can play a positive role and encourage learners' behaviour and ease social changes (Cherry, 2019). This is applicable also to the social learning process of immigrant women.

As immigrant women usually create a safety network that is formed of people in their community who share similar cultural values, a role model from among their peers could demonstrate to other immigrant women a useful social learning path. An experienced mentor can demonstrate new, positive behaviour to encourage learning (EO, 2018–2019; Cherry, 2019)

Given the changing environment of migration, it can be concluded that immigrant women live under conditions of complexity and uncertainty. They need to build the capacity to learn about autonomy and relatedness. Social learning encompasses many factors that can help immigrant women learn about the competitive advantages of enhancing their own capacities in the host society.

Autonomy learning is a learning process that takes place on the level of the individual. Learning about the host country and its traditions can construct logic to accelerate autonomy learning among immigrant women. The development of immigrant women's autonomy can lead to increased willingness to make decisions and to take independent actions regarding different solutions required for integration in the North. Comparative social analysis and more analysis of their learning capabilities in terms of better integration can continually expand women's interest in accommodating different logic to learn about new values, norms, and customs of the host country. However, social relationships facilitate the gaining of information about the relevant gender approach of the host country, which in turn can support the empowerment of immigrant women. Capacity-building demands specific evidence on which actions are functioning and which are not, and with regard to what needs to change in the new environment. This realization marks an important phase in social learning. Relatedness to a new environment is not an individual learning process: peers, whether from a similar culture or from the host country, can facilitate and support transformation. Collective discussion and socially active participation help create trustworthy relationships. New knowledge and concrete good evidence can support and increase the understanding of immigrants about their new environment. Sometimes, the involvement of men in women's social learning process is seen as effective, and this could support family members to understand women's autonomy and understand contingencies of transformation or the reform of the family structure as a consequence of migration. In many immigrant families, having a job might not be an attractive option for immigrant women. Taking an initiative to teach them about the labour market, it is seen as necessary to justify accessing the labour market. Learning about the positives aspect of social participation could increase self-motivation for learning.

Immigrant women often face particular barriers that make it more difficult for them to integrate into a new society than it might be for males who are prioritized in the family background. These barriers are often dependent on the culture of the country of origin. Especially women from cultures in which wage earning by women is not seen as the norm might face obstacles when trying to integrate into Finnish society. Different programmes for healthy relationship development between men and women with immigrant backgrounds could be a potential part of the social learning of immigrants. In addition to learning the language, understanding different societal attitudes and expectations concerning the role and rights of women in Finnish society can go a long way towards contributing to a successful integration into the country's society. The highly egalitarian nature of Finnish society might well provide a significant challenge for many immigrants; it does, however, provide for numerous learning possibilities, in particular for immigrant women from very different cultural backgrounds. It is this learning that allows immigrant women to take an active role in their process of integrating into Finnish society.

References

Adler, N. E., Adashi, E. Y., Agular-Gaxiola, S., Agular-Gaxiola, S., Amaro, H., Anthony, M., Brown, D. R., … Snetselaar, L. G. (2010). Research on determinants of women's mental health. In Committee on Women's Health Research, Institute of Medicine (Ed.), *Women's health research: Progress, pitfalls, and promise* (pp. 35–77). Washington, DC: The National Academies Press.

Ala-Mantila, M., & Fleischmann, F. (2017). Gender differences in labour market integration trajectories of recently arrived migrants in the Netherlands. *Journal of Ethnic and Migration Studies*, 44(11), 1818–1840.

Alfred, M. V. (2013). Transnational migration, social capital and lifelong learning in the USA. In S. Guo (Ed.), *Transnational migration and lifelong learning: global issues and perspectives* (pp. 72–88). Canada: Routledge.

Alivernini, F., & Lucidi, F. (2011). Relationship between social context, self-efficacy, motivation, academic achievement, and intention to drop out of high school: A longitudinal study. *The Journal of Educational Research*, 104(4), 241–252.

Allan, G. (1998). Friendship, sociology and social structure. *Journal of Social and Personal Relationships*, 15(5), 685–702.

Amit, K., & Bar-Lev, S. (2014). Immigrants' sense of belonging to the host country: The role of life satisfaction, language proficiency, and religious motives. *Social Indicators Research*, 124(3), 947–961.

Archer, M. (1995). *Realist social theory*. Cambridge: Cambridge University Press.

Archer, M. (2007). *Making our way through the world*. Cambridge: Cambridge University Press.

Bandura, A. (1977). *Social learning theory*. Englewood Cliffs, NJ: Prentice Hall.

Beardwell, I., Holden, L., & Claydon, T. (2001). *Human resource management: A contemporary approach*. Harlow, UK: Pearson Education.

Bernal, G., & Scharrón-del Río, M. R. (2001). Are empirically supported treatments valid for ethnic minorities? Toward an alternative approach for treatment research. *Cultural Diversity & Ethnic Minority Psychology*, 7(4), 328–342.

Block, L. (2015). Regulating membership: Explaining restriction and stratification of family migration in Europe. *Journal of Family Issues*, 36(11), 1433–1452.

Bowers, A. J., & Sprott, R. (2012). Examining the multiple trajectories associated with dropping out of high school: A growth mixture model analysis. *The Journal of Educational Research*, 105(3), 176–195.

Bretherton, I. (1985). Attachment theory: Retrospect and prospect. *Monographs of the Society for Research in Child Development*, 50(1/2), 3–35.

Cesaroli, C. P., Nicklin, J. M., & Ford, M. T. (2014). Intrinsic motivation and extrinsic incentives jointly predict performance: A 40-year meta-analysis. *Psychological Bulletin*, 140(4), 980–1008.

Chang, C.-C., & Holm, G. (2017). Perceived challenges and barriers to employment: The experiences of university educated Taiwanese women in Finland. In E. Heikkilä (Ed.), *Immigrants and the labour markets* (pp. 1–245). Turku: Migration Institute.

Cherry, K. (2019). How social learning theory works. Retrieved from: www.verywellmind. com/social-learning-theory-2795074#citation-1

Craig, G. (2015). *Migration and integration: A local and experiential perspective*. IRIS Working Paper Series, No. 7/2014. Birmingham: Institute for Research into Superdiversity.

Crittenden, P. M. (1990). Internal representational models of attachment relationships. *Infant Mental Health Journal*, 11(3), 259–277.

Daniélé, J. (2016). Women from Muslim communities: Autonomy and capacity of action. *Sociology*, 51(4), 816–832.

Deci, E. L., & Ryan, R. M. (1985). *Intrinsic motivation and self-determination in human behavior (Perspectives in social psychology)*. New York: Springer.

Deci, E. L., & Ryan, R. M. (2000). The "what" and "why" of goal pursuits: Human needs and the self-determination of behavior. *Psychological Inquiry*, 11(4), 227–268.

Deci, E. L., & Ryan, R. M. (2002). An overview of self-determination theory: An organismic dialectical perspective. In E. L. Deci, & R. M. Ryan (Eds.), *Handbook of self-determination research* (pp. 3–33). Rochester: University of Rochester Press.

Deci, E. L., Ryan, R. M., & Williams, G. C. (1996). Need satisfaction and the self-regulation of learning. *Learning and Individual Differences*, 8(3), 165–183.

Dey, C. (2002). Methodological issues. *Accounting, Auditing & Accountability Journal*, 15(1), 106–121.

Dombrowski, S. U., Knittle, K., Avenell, A., Araujo-Soares, V., & Sniehotta, F. F. (2014). Long term maintenance of weight loss with non-surgical interventions in obese adults: Systematic review and meta-analyses of randomised controlled trials. *BMJ*, 348, g2646.

Donnelly, T. T., Hwang, J. J., Este, D., Ewashen, C., Adair, C., & Clinton, M. (2011). If I was going to kill myself, I wouldn't be calling you. I am asking for help: Challenges influencing immigrant and refugee women's mental health. *Mental Health Nursing*, 32(5), 279–290.

Douki, S., Ben Zineb, S., Nacef, F., & Halbreich, U. (2007). Women's mental health in the Muslim world: Cultural, religious, and social issues. *Journal of Affective Disorders*, 102(1–3), 177–189.

Ducu, V. (2018). Afterword: Gender practices in transnational families. In V. Ducu, M. Nedelcu, & A. Telegdi-Csetri (Eds.), *Childhood and parenting in transnational settings* (vol. 15). (pp. 191–204). Cham: International Perspectives on Migration.

Duncan, J. S., & Lambert, D. (2004). Landscape of home. In J. S. Duncan, N. C. Johnson, & R. H. Schein (Eds.), *A companion of cultural geography* (pp. 382–403). Malden, MA: Blackwell.

Ebrahim, R. (2017). Women's entitlement to autonomy in Islam and related controversies surrounding verse 4: 34. In E. Aslan, & M. Hermansen (Eds.), *Religion and violence. Wiener Beiträge zur Islamforschung*. Wiesbaden: Springer.

EO. (2018). Ethnographic observation 2018. Unpublished material.

EO. (2018–2019). Ethnographic observation between 2018–2019. Unpublished material.

EU. (2018). Integration of immigrant women: A key challenge with limited policy resources. Retrieved on August 8, 2019 at: https://ec.europa.eu/migrant-integration/feature/integration-of-migrant-women

Fall, A. M., & Roberts, G. (2012). High school dropouts: Interactions between social context, self-perceptions, school engagement, and student dropout. *Journal of Adolescence*, 35(4), 787–798.

Forsander, A. (2008). Integration through the Nordic welfare state: Does work make you in to a real Finn? In H. Bloomberg, A. Forsander, C. Kroll, P. Salmeenhaara, & M. Similä (Eds.), *Sameness and diversity*. (pp. 71–91). Finland: Helsinki Research Institute.

Fortier, M. S., Duda, J. L., Guerin, E., & Teixeira, P. J. (2012). Promoting physical activity: Development and testing of self-determination theory-based interventions. *International Journal of Behavioral Nutrition and Physical Activity*, 9, 20.

Fresnoza-Flot, A., & Shinozaki, K. (2017). Transnational perspectives on intersecting experiences: Gender, social class and generation among southeast Asian migrants and their families. *Journal of Ethnic and Migration Studies*, 43(6), 867–884.

Gagnon, J. C., Fortier, M., McFadden, T., & Plante, Y. (2018). Investigating the behaviour change techniques and motivational interviewing techniques in physical activity counselling sessions. *Psychology of Sport & Exercise*, 36, 90–99.

Ghorashi, H. (2017). Negotiating belonging beyond rootedness: Unsettling the sedentary bias in the Dutch culturalist discourse. *Ethnic and Racial Studies*, 40(14), 2426–2443.

GIP (Government Integration Programme) (2016–2019) Government Resolution on a Government Integration Programme, Retrieved on April 27, 2019 at: https://julkaisut.valtioneuvosto.fi/handle/10024/79156

Guo, S. (2013). *Transnational migration and lifelong learning: Global issues and perspectives*. New York: Routledge.

Guruge, S., Kanthasamy, P., & Santos, E. J. (2008). Addressing older women's health: A pressing need. In S. Guruge, & E. Collins (Eds.), *Working with immigrant women: Issues and strategies for mental health professionals*, pp. 235–258. Toronto: Centre for Addiction and Mental Health.

Haan, M. (2008). The place of place: Location and immigrant economic well-being in Canada. *Population Research and Policy Review*, 27(6), 751–771.

Hall, G. N. (2003). Cultural competence in clinical psychology research. *Clinical Psychologist*, 56, 11–16.

Huddle, J. (2018). Advantages & disadvantages of ethnographic research. Retrieved on August 8, 2019 at: https://classroom.synonym.com

James, W. (1890). *The principles of psychology* (vol. 1). New York: Holt.

Johnson, J. (1981). Old values – new directions: Competences, adaptation, integration. *American Journal of Occupational Therapy*, 35(9), 589–598.

Kathryn, R. W., & Miele, D. B. (2016). *A handbook of motivation at school*. New York: Routledge.

Kendra, C. (2017). What is self-determination theory? Retrieved on June 25, 2019 at: www.verywellmind.com/what-is-self-determination-theory-2795387

Kilkey, M., & Palenga-Möllenbeck, E. (Eds.). (2016). *Family life in an age of migration and mobility global perspectives through the life course*. London: Palgrave Macmillan.

Klein, D. (1998). *The strategic management of intellectual capital*. Woburn: Butterworth-Heinemann.

Kolb, D. A., & Fry, R. E. (1974). *Toward an applied theory of experiential learning*. MIT Alfred P. Cambridge, MA:Sloan School of Management.

LeCompte, M. D. (1982). Problems of reliability and validity in ethnographic research. *Review of Educational Research*, 52(1), 31–60.

Lee, Y.-S., & Hadeed, L. (2009). Intimate partner violence among Asian immigrant communities: Health/mental health consequences. *Help-seeking Behaviors, and Service Utilization. Trauma, Violence, and Abuse*, 10(2), 143–170.

Legault, L., Green-Demers, I., & Pelletier, L. G. (2006). Why do high school students lack motivation in the classroom? Toward an understanding of academic amotivation and the role of social support. *Journal of Educational Psychology*, 98(3), 567–582.

Lerner, J., Rappaport, T., & Lomsky-Feder, E. (2007). The ethnic script in action: The regrounding of Russian Jewish immigrants in Israel. *Ethos*, 35(2), 168–195.

Levitt, P. (2003). "You know Abraham was really the first immigrant": Religion and transnational migration. *International Migration Review*, 37(3), 847–873.

Marchetti, S., & Salih, R. (2017). Policing gender mobilities: Interrogating the "feminisation of migration" to Europe. *International Review of Sociology*, 27(1), 6–24.

Marcotte, R. D. (2010). Muslim women in Canada: Autonomy and empowerment. *Journal of Muslim Minority Affairs*, 30(3), 357–373.

Martin, J. (2010). *Key concepts in human resource management*. London: Sage.

Mata, J., Silva, M. N., Vieira, P. N., Carraça, E. V., Andrade, A. M., Coutinho, S. R., & Teixeira, P. J. (2009). Motivational "spill-over" during weight control: Increased self-determination and exercise intrinsic motivation predict eating self-regulation. *Health Psychology*, 28(6), 708–716.

Moore, G. F., Audrey, S., Barker, M., Bond, L., Bonell, C., Hardeman, W., & Wight, D. (2015). Process evaluation of complex interventions: Medical research council guidance. *BMJ*, 350, h1258.

Nelson, M. E., Rejeski, W. J., Blair, S. N., Duncan, P. W., Judge, J. O., King, A. C., Macera, C. A., & Castaneda-Sceppa, C. (2007). Physical activity and public health in older adults: Recommendation from the American College of Sports Medicine and the American Heart Association. *Medicine & Science in Sports & Exercise*, 39(8), 1435–1445.

O'Mahony, J. M., & Donnelly, T. T. (2007). The influence of culture on immigrant women's mental health care experiences from the perspectives of health care providers. *Mental Health Nursing*, 28(5), 453–471.

OECD. (2018). Working together: Skills and labour market integration of immigrants and their children in Finland. Retrieved from: 10.1787/9789264305250-en

Parker, J. G., & Seal, J. (1996). Forming, losing, renewing, and replacing friendships: Applying temporal parameters to the assessment of children's friendship experiences. *Child Development*, 67(5), 2248–2268.

Pelletier, L. G., & Sharp, E. C. (2009). Administrative pressures and teachers' interpersonal behaviour in the classroom. *Theory and Research in Education*, 7(2), 174–183.

Ryan, R. M., & Deci, E. L. (2016). Facilitating and hindering motivation, learning and wellbeing in schools: Research and observations from self-determination theory. In W. R. Kathryn, & D. B. Miele (Eds.), *A handbook of motivation at school* (pp. 96–119). New York: Routledge.

Sarvimäki, M. (2011). Assimilation to a welfare state: Labor market performance and use of social benefits by immigrants to Finland. *The Scandinavian Journal of Economics*, 113(3), 665–688.

Sarvimäki, M., & Hämäläinen, K. (2010). *Assimilating immigrants: The impact of an integration program*. HECER Discussion Paper 306. Helsinki: Helsinki Center of Economic Research (HECER).

Saukkonen, P. (2016). *From fragmentation to integration: Dealing with migration flows in Finland*. Helsinki: Sitra Memos.

Schneebaum, A., Rumplmaier, B., & Altzinger, W. (2015). Gender and migration background in intergenerational educational mobility. *Education Economics*, 24(3), 239–260.

Sigmon, S. T., Whitcomb, S. R., & Snyder, C. R. (2002). Psychological home. In A. T. Fisher, C. C. Sonn, & B. J. Bishop (Eds.), *Psychological sense of community: Research, applications and implications* (pp. 24–41). New York: Kluwer.

Social Planning Council of Ottawa. (2010). *Immigrant children, youth and families: A qualitative analysis of the challenges of integration*. Ottawa: Author.

Spitzer, D. L., Rasouli, M., Hyman, I., Galabuzi, G.-E., Hadi, A., Mercado, R., & Patychuk, D. (2012). Exploring the experiences of socially and economically disadvantaged people in Canada: Qualitative analysis. *Report for Human Resources and Skills Development Canada*.

Su, Y. L., & Reeve, J. (2011). A meta-analysis of the effectiveness of intervention programs designed to support autonomy. *Educational Psychology Review*, 23(1), 159–188.

Sue, S. (1999). Science, ethnicity, and bias: Where have we gone wrong? *American Psychologist*, 54, 1070–1077.

Sue, S. (2003). In defense of cultural competency in psychotherapy and treatment. *American Psychologist*, 58, 964–970.

Teixeira, P. J., Silva, M. N., Mata, J., Palmeira, A. L., & Markland, D. (2012). Motivation, self-determination, and long-term weight control. *International Journal of Behavioral Nutrition and Physical Activity*, 9(22), https://doi.org/10.1186/1479-5868-9-22.

Telegdi-Csetri, Á., & Ducu, V. (2016). Transnational difference – Cosmopolitan meaning. In V. Ducu, & Á. Telegdi-Csetri (Eds.), *Managing difference in Eastern-European transnational families* (pp. 13–27). Frankfurt am Main: Peter Lang.

THL. (2018). Immigrants' health and well-being. Retrieved August 31, 2019 at: https://thl.fi/en/web/migration-and-cultural-diversity/immigrants-health-and-wellbeing

Webb, T. L., Joseph, J., Yardley, L., & Michie, S. (2010). Using the internet to promote health behavior change: A systematic review and meta-analysis of the impact of theoretical basis, use of behavior change techniques, and mode of delivery on efficacy. *Journal of Medical Internet Research*, 12(1), e4.

Wilson, D. K., Griffin, S., Saunders, R. P., Evans, A., Mixon, G., Wright, M., & Freelove, J. (2006). Formative evaluation of a motivational intervention for increasing physical activity in underserved youth. *Evaluation and Program Planning*, 29(3), 260–268.

Wray, H. (2009). Moulding the migrant family. *Legal Studies*, 29(4), 592–618.

Yeasmin, N., & Koivurova, T. (2019). Immigrant women and their social adaptation in the arctic. In S. Uusiautti, & N. Yeasmin (Eds.), *Human migration in the Arctic: Past, present and future* (pp. 67–89). Singapore: Palgrave Macmillan.

6 An integral assessment of relevant perspectives of legal pluralism and the family laws of immigrants

Waliul Hasanat, Nafisa Yeasmin and Timo Koivurova

Introduction

Sharia law and the practising of Muslim family laws (MFL) have been the subject of heated discussion among the Finns in the recent years. Topics include female genital mutilation (FGM), and the demand for Islamic personal law is increasing day by day. In fact, Finland has to receive a certain number of refugees on humanitarian grounds every year. Most of them are from Islamic states, which leads to a debate on legal pluralism. The debate has also been constructing and perpetuating stereotypes, causing misunderstandings, fear, and panic among the Finns as well as the immigrants.

Stereotypical discussions on Sharia law are based on the hidden and invisible practice of a strong culture and social norms. Sharia law has been misinterpreted by many Finns and even by many other non-Muslim immigrants. This situation has led to discussion on legal pluralism (Griffiths, 1986), which refers to a situation in which a state or sovereign power supports different bodies of law for different groups in a society (ibid.). Previous research has stated that strong feelings regarding one's own ethnic culture and beliefs are more positive, and these cultural practices and beliefs may clash with the norms, statutory laws, non-statutory laws, and civil laws of the host country (Springer & Martini, 2015). The members of an ethnic group in a sovereign state follow the domestic laws of the state in which they live and deal with public international law. This practice of different legal systems can cause problems sometimes in conceptualising the domestic laws of the host countries.

Many immigrant families in Finland follow their own family law that is based on the domestic law of their state of origin. Examples include some laws regarding marriage and divorce, along with various modes and values of their community laws such as social welfare, gender relations, and other related rights. As such, within a single host polity, there is the practice of different heterogeneous systems of norms, customs, and community laws, and recognition of these several operations is defined as legal pluralism (Schiller, 2015). Legal pluralism can cause some social dilemmas in the host society to varying degrees. It influences a reshaping of several legal systems of law and power (Ong, 1998) in order to maintain social coherency.

In our study, we explored the impact of legal pluralism and the effect that the family resilience of Muslim immigrants can have on a host country such as Finland.

Recognising new laws, orders, and obligations in the statutory laws of the host country can create confusion in determining a comfortable system that can be either resilient or not among Muslim immigrants. Social-ecological resilience in a new society depends on dynamic mutual adjustments and different challenges (Anderies et al., 2004; Berkes, 2006; Schlüter & Pahl-Wostl, 2007; Armitage, 2008). Muslim immigrants, still a minor group among minorities, have a fear of losing their fundamental identities, which face pressure from various bodies of law in the host country. For example, the practice of FGM recently generated a public debate in Finland that led to a public initiative to ban the practice (Yle, 2018). The public debate can, on the other hand, inspire questions of cultural relativism and incite criticism of legal pluralism. In this study, we discuss an integral pluralist concept to understand the relationship between the subjective degrees of plurality in practising MFL and the collective social systems including care, culture and the nature of individuals and group notions. A systemic integral assessment of relevant perspectives can draw an insight into different postulates for sustaining different versions of legal pluralism. As a foundational theory, we have illustrated Wilber's integral approach, which embeds an integral legal infrastructure that can allow various legal systems. Such an integral approach may influence the adaptation and transformation of plural legal orders so that no autonomous identity (Von Benda-Beckman & Von Benda-Beckman, 2006) can lose its real autonomy. The integral approach explains the recognition of mutual ideologies that advance legal pluralism and can also inhibit the growth of legal pluralism by avoiding the monopolisation of laws (Ottley & Zorn, 1983).

Theoretical framework

Interrelationships between immigrant people and a social ecological system appear as a general, neutral, and essential means for maintaining effective integration. Immigrants can control their integration process according to their rights and characterise their territorial existence based on their essential needs. Both of these factors can directly or indirectly (Yeasmin, 2018) affect the path of their integration into a new society (ibid.). The practice of their own cultural rights along with family rights are important to understanding immigrants' identities in the host country. However, maintaining the host socio-ecological system demands a capacity for transformability in a new domain, which emphasises adaptive resource management (Walker et al., 2004). Integrating new legal concepts and components can cause detrimental clashes in a new domain. Some legal scholars also see the issues as belonging to the group rights of immigrants, rather than an individual's rights. When dealing with Muslim immigrants and their right to assert their religious and family freedoms, it is indeed a particular case of a religious group rather than individuals (Sense about Sharia, 2010). Such rights include: (1) the right to identity and special measures for the preservation of this right; (2) the right of religious freedom to practise one's own religion; (3) the right to social communication with similar religious minorities within or outside the country; 4) the rights to freedom of autonomy (Volkov, 2000; Swenson, 2018) to enjoy family life, which includes child custody, patrimony, marriage contracts, and so on.

On the contrary, immigrant minorities are obliged to follow the legal systems of the host country, which can contradict the above-mentioned rights to a large extent. In these circumstances, non-state legal systems exist simply as a subordinate legal system of the state legal system in the host country (Lerner, 2011a, 2011b).

In Finland, the criticism of legal pluralism, which is incited by multiculturalism, is not particularly stronger than it is in the Netherlands or in the United Kingdom. However, Finnish immigration and integration policies have a spillover effect on the Finnish legislative system such as social policies, human rights, cultural rights, and family rights. Integration policies have a legal political domain along with the socio-economic and cultural domains, which also include the religious domain (Penninx, 2004). The policies create an integral plurality towards integration – "porous legality" (Griffiths, 1986) – for state, and "legal porosity" (ibid.) for Muslim immigrants. They face diasporic legal cultures and a legal system of differences (Shah, 2005) and are compelled to act against the legal system of the host state when it contradicts their own diasporic legal system (Shah, 2005).

A successful integration process needs: (1) the willingness of the state or society to provide the appropriate opportunities for immigrants to integrate and (2) the commitment of immigrants towards the host society (Weiner, 1996). State policies are willing to implant immigrants but without legal transplantation. In contrast, Muslim immigrants are also willing to integrate into the host society, without, however, leaving their diasporic identity in general.

The banning of halal slaughtering and FGM have been the main topics of discussion on the secular absolutistic model, which has been seen as a monopoly of law-making among some immigrant religious groups in Finland (Yeasmin, 2014). Legal pluralism has become a major theme in Finnish socio-legal studies after the banning and denying the religious and family rights of some groups of Muslim immigrants. Practising non-statutory laws can create a gap between legal practice and textual legal provision (Vanderlinden, 1989), because the positive manifestation of the conformity of textual law and raising alternative legal ordering can increase the legitimacy of the state law. The convergence of the classic themes of MFL and cultural rights is challenging and cannot stop the practice of these bodies of law by legal ordering to some extent, which sadly has the effect of making laws and orders inefficient (Roberts, 1979). Clarifying the features of the case of legal pluralism among the target groups practising different legal systems is effective (Woodman, 1998). Encompassing any law into an existing system in order to establish normative orders and human experiences needs interpretation of concepts, exercises, and continuous legal education for porosity (Santos, 1989, 1995; Guevara-Gill & Thome, 1992).

In order to provide an inclusive framework for legal pluralism, we need to separate a meta-paradigm that can support the understanding of the topographies of the plurality of laws and legal practices available in Finland. Wilber's integral theory allows us, here in this study, to systematically explore and develop multiple aspects of legal pluralism. This integral theory relates more of the reality of the need to practise the combined laws and orders together with the phenomenological aspects of individual values in addition to a certain cognitive awareness under various organisational systems and behaviour. Likewise, Muslim immigrants find that the Finnish liberal constitutional democracies lack a certain legal accommodation of their family law and cultural rights.

In this theory, the interior quadrant(s) focus on the reality of the informal practices of the cultural and family rights of Muslim immigrants – individually and collectively; in contrast, the exterior quadrant(s) highlight the statutory laws and institutionalised practices of Finnish law that ban all types of ritualism and ethno-communal practices such as FGM and the dignity or honour of men in the family. This gives us a useful understanding of the complexity of the reality in a way that addresses problems that can be digested in a more skilful way (Esbjörn-Hargens & Zimmerman, 2009; Esbjörn-Hargens, 2018). This assessment also incorporates the essential tracks that are in need of attention.

The integral theory (see Figure 6.1), with its eight methodological zones, is named methodological pluralism and includes the phenomenological approach of subjective realities in the first quadrant, and the ethnomethodological approach, which indicates collective or intersubjective realities on diasporic laws and rights. Quadrant IT is an exploration of the empirical realities of the host state that represent and introduce the most systematic approach for Muslim immigrants by ensuring that they are bound to follow the tried and factual methods of law on FGM, human rights, and other western rights that have been respected by all Muslim immigrants living in Finland.

	INTERIOR	EXTERIOR
I N D I V I D U A L	Structuralism: subjective realities: subjective values and intentionality of Muslim immigrants create substantive constraints on conformity of state law and create a self-regulating family life based on non-state family law and cultural identity	IT: empiricism: inter-objective dynamics of the state is to integrate immigrants by developing strong legal regime(s). Here, normative legal pluralism is dealing with a specific anthropological legal pluralism
C O L L E C T I V E	WE: ethnomethodology: cultural ritualism or group rights and values create an ethical relationship with family and cultural rights	ITs: institutionalism/system theory Dissolving subjective and intersubjective realisms into their objective aspects. Legal centralism and a hierarchy of laws that could create either weak or strong conceptions of legal pluralism

Figure 6.1 Combining Wilber's integral theory in practice with subject matter

Source: Wilber's integral theory (Wilber, 1996)

This methodological pluralism has three pillars: (1) inclusion – consulting multiple laws and orders, (2) enfoldment – prioritising the importance of laws and orders, and (3) enactment – recognising or believing that the laws and orders are created after consultation with the target groups, and that these laws and orders are known to everybody. This methodological pluralism includes major and valid insights into understanding the concept of legal pluralism in a comprehensive way. This model gives us an effective response to the complexity of reality on normative legal pluralism.

Methodology

In order to encourage a discussion on the varieties of legal pluralism, we organised three different focus group discussions, each of 2 hours' duration. These three group discussions attracted 30 immigrants and Finns. The immigrants were from different religious backgrounds. The first group discussion (G1) attracted more Finnish women than immigrant men and women. We continued our discussion with 15 participants, 6 of whom were immigrants (1 man + 5 women), and 9 were Finns. The immigrants were from Pakistan, Uzbekistan, Afghanistan, and Palestine. The second group discussion (G2) was organised with four immigrant women from Uzbekistan, Thailand, Nigeria, and Russia, and one male participant from Syria. Immigrants from Palestine, Syria, Iraq, Pakistan, Brazil, Algeria, Greece, Lithuania, and Russia participated in the third discussion group (G3). The third group consisted of four men and six women.

Our goal was to discuss four aspects that overlap between statutory law and diasporic beliefs: (1) religious or diasporic particularism, (2) the informal practice of non-statutory laws, (3) the formalised jurisdiction on diasporic law, and (4) the legal conflicts regime. The formal practice of non-statutory laws is usually invisible. Some practices are visible, but the majority remain highly invisible within Finnish society, and even within their own communities. One of our discussion sessions had well-informed participants; however, two of the discussions were semi-structured, and the respondents were not informed beforehand about the discussion session.

At the beginning of the session, we informed the participants of the thematic content of the discussion. We explicitly operated the discussion in the shadow of Sharia law and in parallel with the statutory laws of Finland. Our goals were to monitor whether our respondents are well informed about the jurisdiction of the statutory laws to enforce the Sharia law, and to make possible a comparison of the fundamental rights of immigrant Muslim minorities with other legal regimes of statutory law and jurisdictions in Finland.

We tried to keep our discussion focused on the equal rights of women and girls in Muslim families and whether those rights are somehow assured by MFL or community law of their country of origin, and we scrutinised how statutory laws have been monitored and controlled in their family life in the host region. The discussion gave us a chance for convergence between statutory law and diasporic community law; also, the discussion supported our study to reveal how the convergence and execution of Finnish statutory law and MFL shape a debate on legal pluralism. The respondent-led discussion also partly focused on the debate about the support measures if any of the family rights overrides the domestic law of Finland.

Although our discussion covered many other side-effects of social control and regulation undermining the legal practices of statutory law alongside co-ordination of behaviour of some participants of the target group, the respondents' holistic cultural pluralism was discussed in the group. They agreed that nationwide discussion on cultural pluralism could foster a radical theory of legal pluralism in Finland.

Based on our discussions and the theoretical framework, we have set four hypotheses, as seen in Figure 6.2.

Description

In 2018, Finland took the initiative to ban FGM (Yle, 2018) and place emphasis on the rights of immigrant women. Many members of this focus group still lacked vital information about this related initiative of laws. Moreover, the OECD report (OECD, 2018) also states that immigrant women are at risk of marginalisation in Finnish society, as, generally, their integration path ends with home-care benefits. The integration of immigrant women has remained a subject of discussion.

Our discussion led to four hypotheses.

	INTERIOR	EXTERIOR
I N D I V I D U A L	Structuralism: Hypothesis 1: subjective realities, e.g. family laws, individual honour, values and cultural rights inhibit social changes in Finland	IT: empiricism: Hypothesis 2: inter-objective dynamics and the target of state laws are to enable social changes
C O L L E C T I V E	WE: ethnomethodology: Hypothesis 3: collective ethical, cultural, and religious rights create a radical theory on legal pluralism	ITs: Institutionalism/system theory: Hypothesis 4: state laws ensure systems for socio-ecological resilience and set shared values

Figure 6.2 Hypothesised Wilber's theory in practice

Hypothesis 1: subjective realities – for example, family laws, individual honour, values, and cultural rights – inhibit social change in Finland

Fundamental rights in Finland include that all citizens have an equal right to practise their religion, which has created liberal-democratic institutional pluralism that accommodates MFL in Finland. To a greater or lesser extent, all immigrant women, including those belonging to Muslim families from Iraq, Palestine, Afghanistan, or Somalia, have faced unfair choices between their rights, culture, and religion. Families are afforded far-reaching autonomy to choose that women spend their time and energy taking care of the home rather than taking advantage of integration training, which excludes immigrant women from the core Finnish society (Finnish respondents). Some Finnish respondents believe that Finland needs specific practices of regime with jurisdiction based on the concept of equal respect, as many Finns believe that members of Muslim immigrant families enjoy rights recognised under MFL, although not covered under Finnish statutory law, when women's and children's rights/interests are not internally ensured. Effective control over these type of MFL by the Finnish legal system can be more effective in solving the dilemma.

Some immigrant men expressed the opposite view: immigrant women are habituated by the family rights and cultural rights of their country of origin, and sometimes it is their own decision to choose their path as a housewife after immigrating to Finland, and the decisions are controlled by immigrant women only (immigrant men from Iraq, 2018 group discussion).

However, the third discussion group discussed the autonomous epistemological subject of bad faith and the imbalance of power towards men. Men hold superior authority in the family and decide all family matters alone on behalf of the family (immigrant women, 2018 group discussion). The core patriarchal structure of the family practice of many religious groups also inhibits social change in Finland.

Many families also create a self-regulating social rule on FGM, as it is restricted by the host country but not, however, restricted by their country of origin. Therefore, they send their daughters to their home country, a practice that they find to be a safe and permitted social rule as a phenomenological counterpart to legal pluralism (Finnish respondents, 2018 group discussion).

Many immigrants find that the practice of their own identity and rights supports them to acculturate into Finnish society, and, therefore, sometimes Finnish core values support the harmonisation of laws to encourage immigrants to act in a consistent manner with Finnish law. However, owing to a lack of awareness, immigrants feel little or no obligation to follow the law, which empowers their sense of autonomy and independent legitimacy, and this in turn inhibits social change (immigrant women from Europe, 2018 group discussion).

The practice of discrimination against women is a frequent phenomenon in many immigrant families, and sometimes immigrants with a Finnish spouse also are discriminated against, but usually those practices are seen as a matter of accepted practice (Campbell & Swenson, 2016) rather than a violation of state law (immigrant women, 2018 group discussion).

Many of the respondents believe that FGM practice, along with other practices of MFL or family laws of other immigrants, needs to be discouraged, and those who engage in such practices need to be aware that, in Finland, they are hostile and discriminatory against women.

According to the philosophy of human movement, its objective is to investigate humans' identity under different phenomena as they reportedly experience and perceive them (Spiegelberg, 1975). The general assessment of legal doctrines reveals that some liabilities are created from negligence (Morano, 1974). Morano provides three dilemmas of negligence, one of which is crucial and interrelated to our study. It states that people need some necessary liabilities otherwise they can suffer from moral condemnation (ibid.). The overall feeling of immigrant respondents is that Finnish society is negligent, as they are treated as a separate group of people in the society (respondents in Group 1 (G1), 2018 group discussion). Although they all have access to the same facilities and receive equal welfare benefits in Finland, however, still they lack some necessary liabilities in the society. In their country of origin, they used to have particular responsibilities to resolve certain societal problems, but Finnish society does not expect such liabilities from them. In Finnish society, this has been leading towards a trend to oppose empiricism to some extent, and immigrants instead rely on utilising their own rights and laws as much as they can. Morano mentioned it as "culpable inadvertence" (Morano, 1974), which explains that Muslim immigrants practise their family laws and cultural rights without question and with only a marginal awareness of statutory laws. In this situation, they can only feel their own concerns and access their personal memories, imagination, and anticipations (Morano, 1974; Hermann, 1982).

Hypothesis 2: the inter-objective dynamics and target of state laws are to enable social changes

The target of rule of law is twofold: to ensure the securities and capacities of the function of the law. To avoid overlapping, it is imperative that attention is paid to the action-guiding function of rule of law, as some gender rights are already secured by the provisions of public international law – for example, the rights of women. Ensuring proper capacity building to enforce existing laws instead of taking any further constructive action could avoid overlapping in legal pluralism (G1, 2018 group discussion). Human rights arose from the concept of how humans feel about their individual rights. We have to draw attention to that rather than paying attention to constantly making rule of law (Group 3 (G3), 2018 group discussion).

Rainer (Frost, 1999) states:

> Hence it is neither very difficult nor unjustified to draw attention to and emphasise the specific genesis of this concept, considering how differently other traditions and cultures understand the meaning of the term "human being".

Finnish national laws provide legal positivism among marginalised groups such as Muslim immigrants and ensure their security. In this case, the Finnish laws deflect other transformative accommodations of statutory laws that could not only facilitate respect for religious exercise, but also could prioritise gender equality in general without establishing or adopting new law that goes against any religious rights (Cohen, 2012). A lack of legal accommodation also causes unfair treatment of some minorities. A strong regime requires a fundamental assumption for constructing a functionalist theory of law (Tamamaha, 1997); because of the lack of sociological relevance, re-specification of some laws can lead to their plural nature (Dupret, 2007).

The rethinking of legal pluralism by legal institutions has a positive impact on society as a whole. According to one of the respondents, before taking any initiative to establish any law, it is significantly important to talk with a focus group and create awareness. Again, according to the respondent, a "law against FGM has been adopted to stop the act, but the act should not be stopped before creating any awareness against FGM". Some of the perspectives of statutory laws are based on western culture, which is rather different from Islamic culture. Therefore, national laws should support the multicultural political reality rather than monocultural social constructivism (Frost, 1999). Many of the respondents agreed that engaging a focus group before making any laws is more effective than involving them later. Many respondents did not even know about the whole Finnish initiative on FGM, and many also lack accurate knowledge of the health problems and human rights aspects of FGM. Therefore, sharing knowledge is important before any laws are adopted. Statutory laws can enable social change through creating awareness among focus groups, although it is still a challenge at a grassroots level (immigrant women in G3, 2018 group discussion). The ethical feelings, health, and well-being of individuals require mutual conceptions of the statutory laws (ibid.; Frost, 1999). The contingencies in enforcing rule of law are mutual awareness, explaining the problems, engaging in debates, and synthesising legal and social regimes. Before synthesising a law, safeguarding must be ensured through acknowledgement and appreciation of laws among the target group (G1, 2018 group discussion). Legal empiricism is contingent on empirical practices, and, therefore, tangible ideologies and sharing understanding are valuable in the case of legal pluralism. Understanding the needs of the target group, having up-to-date information about social changes, and anticipating the effect of those social changes conform to the guidelines of legal empiricism (G1, 2018 group discussion). Although every legal construction has ethical social grounds (Rawls, 1999), the reality of the social structure of the target group should be considered to ensure that the new laws are able to bring about social changes. All of the immigrant respondents in the discussion groups lacked knowledge of the Finnish FGM-banning initiative, and only a few Finnish respondents from G3 were aware of the initiative, which reflects either a level of ignorance on the part of immigrants about the legal constructivism of the host country or a lack of information exchange or lack of the right path to snowball social cohesion.

Hypothesis 3: collective ethical, cultural, and religious rights create a radical theory on legal pluralism

The rights of immigrants regarding family affairs (such as Orthodox, Christian, Muslim) are subject to a rapidly pluralising society that is in contrast to the practice of western family values. States, such as Finland, that are newly receiving immigrants, are extremely organised in their socio-legal perspectives, and are particularly based on a precise legal system are facing challenges in determining immigration policies. A wrong policy can be turned into gross negligence, and immigration policies related to socio-economics, culture, and politics lack some measures for better integration, which can affect immigrants' orientation and practice of host cultural norms and values.

In accordance with previous research (Yeasmin, 2017: 56),

> The sociology of religion is the most significant factor that influences whether immigrants are integrated into the host society. Religion can become more and more important in a phase of an individual's life when he or she faces great social change in a host country.
>
> (see also Berry, 1997; Baumann, 2002; Borup & Ahlin, 2011; Pace, 2014; female immigrant, 2016)

When people lack biological and psychological resources, they tend to become concerned about their identity, which then tends to turn into a social need related to religious and cultural practice (Maslow, 1943). Ultimately, these needs manifest themselves in action and bring new challenges in the practice of cultural and family rights (Yeasmin, 2017). Immigrants try to adjust to the socio-legal environment of the host country; "however, low levels of interaction with the majority society, confusion and misunderstandings, and policy gaps over uncertainty in the labour market create social pressure for immigrants to try to find their identities and individual beliefs during this transition period" (Yeasmin, 2017: 52). During this phase, individuals need the feeling of belonging to a certain group and will then accept certain behaviours that conform to the norms of that group (Yeasmin, 2017). Afterwards, individual and family concerns become a collective concern of negligence. An unambiguous moral tradition to keep their own ethical, cultural, and family rights has engendered intentional liability among them. There is indeed a subconscious willingness to follow one's own culture and cultural rights in order to maintain a collective relationship with one's own community members (Group 2 (G2), 2018 group discussion). These individuals are afraid of not being accepted by their own community, which pushes them to follow their own cultural rights, and this is their only liability in a host country for surviving socially (G2, 2018 group discussion). This traditional legal reasoning hinders the exercise of statutory laws among the group (G2, 2018 group discussion).

In fact, the stigmatisation of certain misdeeds by Muslims in the world creates a traditional reasoning of group negligence towards Muslim immigrants, which also contradicts group concepts, and the ensuing feelings create an unusually complex phenomenon. Such negligence towards a group of Muslim immigrants results in isolation of the group, as stereotyping leads to a radical theory of divergence between the locals and Muslim immigrants (G3, 2018 group discussion). Such stereotyping generates public ignorance towards Muslim immigrants, which indeed leads to socio-economic disadvantages and political ignorance regarding group identity. It can give some space for political opportunists to create negative discourses towards groups that exercise unrecognised rights under a certain legal regime. Such civilisational clashes hinder the possibility of the harmonious integration of minorities (Huntington, 1996; Lewis, 2003; G3 & G1, 2018 group discussion). This radical theory of divergence implies the risks of legal pluralism. Preserving group rights is seen as establishing equality and obtaining a level of freedom of religion practices. On the other hand, transnational communication with similar groups and families also comes as a right to establish and manage one's own social institutions (Lerner, 2011a, 2011b). To a large extent, creating one's own social institutions can cause self-exclusion and an enclave culture that follows its own diasporic laws and the invisible practices of many diasporic laws such as FGM (G3, 2018 group discussion).

The collective practice of ethical, cultural, and religious practices make Muslim immigrants a poorer immigrant group who suffer from greater unemployment and unequal treatment in western societies (Gurr, 1970; Stewart, 2008; Yeasmin, 2018; G3, 2018 group discussion). The practices of their group rights have been seen as a threat to western society (G2, 2018 group discussion), which is the main reasoning that incites hatred towards groups of Muslim immigrants. The perceived threat inspires radical politics towards Muslim immigrants and also hinders the integration of Muslim immigrants into the host society (G2 & G1, 2018 group discussion). Moreover, these attitudes of western politics engender Muslim diasporas by stigmatising all Muslims and placing them in the same box. This stigmatisation of their identities (although Muslim cultural rights vary across Muslim countries) confuses the definition of cultural rights and religious rights, which are a heterogeneous set of rules and orders based on the laws of the country of origin. Bangladesh, Malaysia, Turkey, and Saudi Arabia are Muslim majority states, each with its own diverse set of rules and statutory laws.

Attempting to assign a similar, monolithic religious identity to all Muslim groups (Murshed & Pavan, 2009) is inherently likely to cause radicalisation by treating them as a homogeneous group. This kind of bias can create a radical group of all Muslim immigrants who are dissatisfied with their political treatment by the host (G1 & G3, 2018 group discussion). Every person has a certain level of tolerance towards their religious beliefs and cultural and family rights. The collective grievances of the host country exclude Muslim immigrants from the majority, which can also cause a collective sense of perpetuating the practice of family rights as a result of negligence of majorities (G3, 2018 group discussion; Morano, 1974).

Hypothesis 4: state law ensures systems for socio-ecological resilience and sets shared values

The legal institutionalism of normative pluralism means a relationship between norms and obligations. Norms that incorporate uncommon standards for majority populations need concrete patterns that can create obligations. The reciprocal expectation of the state is to set shared values, beliefs, and modes of conduct based on the reality beyond individuals (Durkheim, 1984). A responsible decision made by a sovereign state will stabilise its order by implementing the clauses with specific guidelines and interpretations. Statutory laws should try to give equal weight and respect to every community and to preserve a framework for resilience-based governance systems. The target of statutory laws is to foster adaptability of immigrants in order to enhance their potential for transformability (Folke et al., 2010).

Individual rights or family rights are a normative choice, and an individual can practise these rights until they no longer create any public interest or responsibilities (G1, 2018 group discussion). In the case of FGM or the normative choice to practise FGM, it is a subject of public debate and is contradictory to human rights, and it needs to be regarded as a legitimate aspect of ensuring the human rights of immigrant women (G1, 2018 group discussion). Such recognition by a state for more precisely managing legal plurality leads to harmonic legal integration, which can then determine a system to be resilient (Pieraccini, 2013).

Immigrants need to focus on a mutual constitution when they are in new socio-ecological systems. A challenge for the state is the presence of immigrants in a new socio-ecological system, and a mutual constitution can support interpersonal relationships; "the argument is that the identity, background, types or kinds of immigrants are more insignificant than the territory, and significantly all belonging to the same territory is the main argument" (Yeasmin, 2018). "In this case, they can choose any of the components from the social ecology box, which are sufficiently accessible for immigrants and can persuade or manipulate immigrants to become integrated into the host region" (Yeasmin, 2018). At this stage of integration, legal resilience or adopting to and respecting the legal system of the host state all support the stability of the socio-ecological system in a new domain. The sociology of statutory laws combines different rules and orders as well as explaining that perhaps not all legal pluralism (systems) fits with certain perspectives; it also explains that not all legal orders are integral – they can be porous. Many of the family rights or cultural rights of immigrants can be porous and might need better legal explanations to correct their misunderstandings (de Santos, 2006). Understanding the explanation of statutory laws influences the level of resilience among a certain group of people. Immigrants should utilise the positivism of state legal institutionalism in order to influence their resilience capacity within a socio-ecological system (G1, 2018 group discussion).

Legal pluralism gives immigrants space to compare different legal systems and their relationships with different social spheres. It might be a case of making changes to grow up well in spite of hardship (Masten, 2001; Theron et al., 2011). Statutory laws for preventing FGM can enable and encourage the dignity of immigrant girls and women, promote their resilience capacity by increasing adaptive behaviour (Masten & Powell, 2003), and also support them during migratory transitions. Immigrants and their philosophy of family rights are acknowledged, but the state responsibility is still to secure shared values and legal cohesion among its people (G3, 2018 group discussion).

Collective resilience encourages individual resilience, and collective sense-making relies on legal orders to some extent, especially when it creates public debate. Individuals always replicate their diasporic laws and, in a certain manner, they also need to follow diasporic laws, as described in Hypothesis 3 in this article. Therefore, certain institutional prohibitions can reduce collective practices of diasporic laws that contradict the laws of the host state and often create conflicting situations. This contradiction has reciprocal topographies to sustain a collective sense of autonomy (G1 & G3, 2018 group discussion). Statutory laws deal with public stress, thereby sustaining the stress as an external interference (Ebbesson & Hey, 2013).

Conclusion

Theorising legal pluralism in the context of incorporating normative choices of non-state norms of Muslim immigrants not only depends on the recognition or rejection of statutory laws and orders, but also depends on the numbers of separate, interconnected paradigms (Wilber, 1996). The upper left quadrant HP1 rationalises that the individual feelings and sensations of Muslim immigrants about practising their family rights depend on subjective realities and state consciousness and respect for their family laws, individual honour, values, and cultural rights. This respect determines their motivation to maintain their familiar realities by practising MFL, which can inhibit social changes or hinder their integration level in Finland. On the other hand, statutory laws are permanent milestones of development in the host countries, and all Muslim immigrants or individuals have permanent access to integrate themselves into the host society by respecting and following its laws, which can ease their integration. However, immigrants lack true conceptions of what lies behind the Finnish national laws, which is an organisational level of complexity that includes different measures for implementing laws and orders among minorities. Many phenomenological aspects are yet to be targeted to enable the state to introduce legal pluralism (sovereign commands).

Quadrant 3 of HP3 articulates the identity of the majority as well as minorities, where both parties are considered when incorporating legal pluralism at national level. According to Griffith (1986), strong pluralism articulates a hierarchical form of the inexorability of statutory laws, which is the priority and relevant in contrast to weak pluralism. In the weak form of pluralism, statutory laws are no more than a legal doctrine. However, the sociology of the different normative laws creates extended feelings of "we" and "you" or "ethnocentric" and moves to establish strong pluralism, in which other non-statutory laws are no more than legal doctrines or can mislead legal centralism by including hierarchical normative orderings.

Returning to the stages of legal institutionalism and exploring legal pluralism from a new angle requires an understanding of the world-centric progressive view (Wilber, 1996), which entails several moral realisations and the moral development of fundamental rungs. There is a need to unfold higher, deeper, stronger, and weaker potentials as discussed in the different hypotheses. This integral theory helps with understanding that upgrading any statutory law or attempting to upgrade any family laws (Lerner, 2003) cannot restrict the use of community laws and rights; the voluntary involvement of the group members, imposing some liabilities on the group, and clarifying the philosophical ideologies of legal postulates among the group can illustrate the discrepancy with legal pluralism.

Legal pluralism in this context is integral with a certain degree of autonomy that incorporates (Evans, 1995) a diverse set of concrete ties that can bind several subjective and objective states, stages, lines, and types of quadrants together, reportedly with each other for continual negotiations of rule of law. Providing information kits on the topics that cause public debate and fostering knowledge of the topics in different languages can be supportive measures for enabling the legislation in practice. According to our research, such an initiative can define legislation through social norms and ideas rather than through inflexible objectives and will improve real conditions as well (G3, 2018 group discussion; Barkin, 2003).

The subjective degrees of plurality in practising family law by Muslim immigrants depend on time, location, and the situation they face in the host state regime. The social dimension indicates that the behaviours of the social group, the shared values and feelings of social groups can strengthen or weaken the degrees of plurality. The evolution of legal pluralism is not a matter of which of them is wrong or right; rather, it is a matter of fact that needs to be adequately understood in this given situation. Legal pluralism refers to the collective social systems, including care, culture, and the nature of individuals and group notions to be refined.

References

Anderies, J. M., Janssen, M. A., & Ostrom, E. (2004). A framework to analyze the robustness of social-ecological systems from an institutional perspective. *Ecology and Society*, 9(1), 18.

Armitage, D. (2008). Governance and the commons in a multi-level world. *International Journal of the Commons*, 2(1), 7–32.

Barkin, J. S. (2003). Realist constructivism. *International Studies Review*, 5(3), 325–342.

Baumann, M. (2002). Migrant settlement, religion and phases of diaspora. *Migration: A European Journal of International Migration and Ethnic Relations*, 33/34/35, 93–117.

Berkes, F. (2006). From community-based resource management to complex systems. *Ecology and Society*, 11(1), 45.

Berkes, F., Colding, J., & Folke, C. (2003). *Navigating Social–Ecological Systems: Building Resilience for Complexity and Change*. Cambridge: Cambridge University Press.

Berry, J. W. (1997). Immigration, acculturation, and adaption. *Applied Psychology: An International Review*, 46(1), 5–34.

Borup, J. & Ahlin, L. (2011). Religion and cultural integration: Vietnamese Catholics and Buddhists in Denmark. *The Journal of Nordic Migration Research*, 1(3), 176–184. doi:10.2478/v10202-011-0015-z

Campbell, M. & Swenson, G. (2016). Legal pluralism and women's rights after conflict: the role of CEDAW. *Columbia Human Rights Law Review*, 48(1), 112–146.

Cohen, J. L. (2012). The politics and risks of the new legal pluralism in the domain of intimacy. *International Journal of Constitutional Law*, 10(2), 380–397.

de Santos, S. B. (2006). The heterogeneous state and legal pluralism in Mozambique. *Law & Society Review*, 40(1), 39–75.

Dupret, B., (2007). Legal pluralism, plurality of laws, and legal practices: theories, critiques, and praxiological re-specification. *European Journal of Legal Studies*, 1(1), 1–26.

Durkheim, É. (1984). *The Rules of Sociological Method*. New York: Free Press.

Ebbesson, J. & Hey, E. (2013). Introduction: where in law is social-ecological resilience? *Ecology and Society*, 18(3), 25.

Esbjörn-Hargens, S. (2018). An overview of integral theory. an all-inclusive framework for the 21st century. Retrieved on August 31, 2019 at: https://dao-foundation.atlassian. net/wiki/spaces/TEALUA/pages/271351851/AN±OVERVIEW±OF±INTEGRAL± THEORY±An±All-Inclusive±Framework±for±the±21st±Century

Esbjörn-Hargens, S. & Zimmerman, M. E. (2009). *Integral Ecology: Uniting Multiple Perspectives on the Natural World.* New York: Shambhala.

Evans, P. B. (1995). *Embedded Autonomy: States and Industrial Transformation.* Princeton, NJ: Princeton University Press.

Folke, C., Carpenter, S. R., Walker, B. H., Scheffer, M., Chapin, F. S., III, & Rockström, J. (2010). Resilience thinking: integrating resilience, adaptability and transformability. *Ecology and Society*, 15(4), 20.

Frost, R. (1999). The basic right to justification: toward a constructive conception of human rights. *Constellations*, 6(1), 35–60.

Griffiths, J. (1986). What is legal pluralism? *Journal of Legal Pluralism and Unofficial Law*, 18(24), 1–55.

Guevara-Gill, A. & Thome, J. (1992). Notes on legal pluralism. *Beyond Law*, 5, 75–102.

Gurr, T. R. (1970). *Why Men Rebel.* Princeton, NJ: Princeton University Press.

Hermann, D. H. J. (1982). Phenomenology, structuralism, hermeneutics, and legal study: applications of contemporary continental ought to legal phenomena. *University of Miami Law Review*, 36(3), 379–410.

Huntington, S. P. (1996). *The Clash of Civilizations and the Remaking of the World Order.* New York: Simon & Schuster.

Lerner, N. (2003). *Group Rights and Discrimination in International Law* (2nd ed.). The Hague: Brill.

Lerner, N. (2011a). Religion and freedom of association. In J. Witte, Jr. & M. Christian Green (Eds.) *Religion and Human Rights: An Introduction.* (pp. 71–87). Oxford University Press, Oxford.

Lerner, N. (2011b). Group rights and legal pluralism. *Emory International Law Review*, 25, 829–851.

Lewis, B. (2003) *The Crisis of Islam: Holy War and Unholy Terror.* London: Weidenfeld & Nicolson.

Maslow, A. H. (1943). A theory of human motivation. *Psychological Review*, 50(4), 370–396. doi:10.1037/h0054346

Masten, A. S. (2001). Ordinary magic: resilience processes in development. *American Psychologist*, 56(3), 227–238.

Masten, A. S. & Powell, J. L. (2003). A resilience framework for research, policy and practice. In S. S. Luthar (Ed.) *Resilience and Vulnerability: Adaptation in the Context of Childhood Adversities* (pp. 1–25). New York: Cambridge University Press.

Morano, D. V. (1974). Phenomenology of negligence. *Journal of British Society of Phenomenology* 5(2), 135–143.

Murshed, S. M. & Pavan, S. (2009). Identity and Islamic radicalization in Western Europe. Economics of Security Working Paper 14. Berlin: Economics of Security.

OECD. (2018). *Working Together: Skills and Labour Market Integration of Immigrants and Their Children in Finland.* doi:10.1787/9789264305250-en

Ong, A. (1998). *Flexible Citizenship: The Cultural Logics of Transnationality.* Durham, NC: Duke University Press.

Ottley, B. L. & Zorn, J. G. (1983). Criminal law in Papua New Guinea: code, custom and the courts in conflict. *The American Journal of Comparative Law*, 31(2), 251–300.

Pace, E. (2014). Religion in motion: migration, religion and social theory. In H. Vilaça, E. Pace, I. Furseth, & P. Pe Ersson, (Eds.) *Changing Soul of Europe: Religions and Migration in Northern and Southern Europe* (pp. 16–18). Routledge, London.

Penninx, R. (2004). Integration processes of migrants in the European Union and policies relating to integration. *Presentation for the Conference on Population Challenges*, 1–19.

Pieraccini, M. (2013). A politicized, legal pluralist analysis of the commons' resilience: the case of the Regoled'Ampezzo. *Ecology and Society*, 18(1), 4.

Rawls, J. (1999). Themes in Kant's moral philosophy. In S. Freeman (Ed.) *Collected Papers*, (pp. 497–528) Cambridge, MA: Harvard University Press.

Roberts, S. (1979). *Order and Dispute: An Introduction to Legal Anthropology*. Harmondsworth: Penguin.

Santos, B. S. (1989). Law: a map of misreading. Toward a postmodern conception of law. *Journal of Law and Society*, 14(3), 279–302.

Santos, B. S. (1995). *Toward a New Common Sense: Law, Science and Politics in the Paradigmatic Transition*, New York: Routledge.

Schiller, N. G. (2015). Trans-border citizenship: an outcome of legal pluralism within transnational social fields. Theory and research in comparative social analysis. Retrieved on August 31, 2019 at: https://cloudfront.escholarship.org/dist/prd/content/qt76j9p6nz/qt76j9p6nz.pdf

Schlüter, M. & Pahl-Wostl, C. (2007). Mechanisms of resilience in common-pool resource management systems: an agent-based model of water use in a river basin. *Ecology and Society*, 12(2), 4.

Sense about Sharia (2010). Islamic law and democracy: Sense about Sharia. *The Economist*, 14 October. Retrieved 27 June, 2017 from www.economist.com/leaders/2010/10/14/sense-about-sharia

Shah, P. (2005). *Legal Pluralism in Conflict: Coping with Cultural Diversity in Law* (1st ed.). London: Glass House Press.

Spiegelberg, H. (1975). *Doing Phenomenology: Essays on and in Phenomenology*. The Hague: Brill.

Springer, V. A. & Martini, P. J. (2015). Sharia in the everyday life of Muslims. Sharia culture and legal pluralism symposium, Retrieved on July, 27 2019 at: www.westernsydney.edu.au/__data/assets/pdf_file/0006/966291/RSRC0079_Legal_Pluralism_Program_v05.pdf

Stewart, F. (2008). Global aspects and implications of horizontal inequalities (HIs): inequalities experienced by Muslims worldwide. *CRISE Working Paper No. 60*. Centre for Research on Inequality, Human Security and Ethnicity.

Swenson, G. (2018). Legal pluralism in theory and practice. *International Studies Review*, 20(3),438–462.

Tamamaha, B. (1997). *Realistic Socio-Legal Theory: Pragmatism and a Social Theory of Law*. Oxford: Oxford Scholarship Online.

Theron, L., Cameron, C. A., Didkowsky, N., Lau, C., Leiberberg, L., & Ungar, M. (2011). A "day in the lives" of four resilient youths: cultural roots of resilience. *Youth & Society*, 43(3), 799–818.

Vanderlinden, J. (1989). Return to legal pluralism: twenty years later. *Journal of Legal Pluralism and Unofficial Law*, 21(28), 149–157.

Volkov, V. (2000). The political economy of protection rackets in the past and the present. *Social Research*, 67(3), 709–744.

Von Benda-Beckman, F. & Von Benda-Beckman, K. (2006). The dynamics of change and continuity in plural legal orders. *The Journal of Legal Pluralism*, 38(53–54), 1–44.

Walker, B., Holling, C., Carpenter, C., & Kinzig, A. (2004). Resilience, adaptability and transformability in social–ecological systems. *Ecology and Society*, 9(2), 5.

Weiner, M. (1996). Determinants of immigrant integration. In N. Carmon (Ed.) *Immigration and Integration in Post Industrial Societies* (pp. 46–62). London: Macmillan Press.

Wilber, K. (1996). *A Brief History of Everything*. Boston and London: Shambhala.

Woodman, G. (1998). Ideological combat and social observation: recent debate about legal pluralism. *Journal of Legal Pluralism and Unofficial Law*, 30(42), 21–59.

Yeasmin, N. (2014). Maahanmuuttajien kotoutuminen Pohjois-Suomessa [Integration of immigrants in Northern Finland]. Report prepared for the Advisory Board for Ethnic Relations, Finnish Ministry of Interior.

Yeasmin, N. (2017). Cultural identities in sustaining religious communities in the Arctic region: an ethnographic analysis of religiosity from the Northern view point. *Journal of Ethnology and Folkloristics*, 11(2), 51–67.

Yeasmin, N. (2018). *The Governance of Immigration Manifests Itself in Those Who are Being Governed: Economic Integration of Immigrants in Arctic Perspectives*. Rovaniemi: Lapland University Press.

Yle. (2018). Citizens' initiative to ban FGM set for parliamentary consideration. Retrieved on July 28, 2019 at: https://yle.fi/uutiset/osasto/news/citizens_initiative_to_ban_fgm_set_for_parliamentary_consideration/10359602

7 Living in nowhere

Juha Suoranta and Robert FitzSimmons

Introduction

At a reception centre in the Finnish countryside, the landscape separates the centre from the other buildings, while a white flag with a red cross flutters near the top of the flagpole. The Finnish Red Cross (along with local municipalities) assists the authorities in housing asylum seekers: it establishes and maintains reception centres in various parts of the country. The organization's international emblem, the white flag with a red cross, is familiar to emergency areas and is used for catastrophic images in magazines and in news broadcasts. The flag does not seem to fit into the peaceful image of the Finnish countryside. But then again, many people who have escaped conflicts, persecution, war, and hunger have found shelter within Finnish country landscapes. And here, in the northern corner of the Earth, they wish to be safe, at least for a moment, under the protection of such a flag.

In this chapter, we first ponder, based on an empirical study (see Suoranta, 2011), how asylum seekers are treated in Finnish reception centres and what problems they experience. Then, based on their experiences and insights, we offer concrete ideas on how life in the centres could be improved to create a more inclusive and hospitable atmosphere.

The year 2015 became a game changer in Finland. In just a few short months, from September to December, thousands of migrants came to Finland through Sweden and from Russia. During that year, Finland received an unprecedented number of applications for international protection – 32,000 in total.[1] Many Finns felt that the country was faced with its own migrant crisis, and that the sudden situation was compromising basic welfare structures, such as the health care system, as well as people's general safety. In the media and in public discussion, the situation was framed as a natural phenomenon and described using such expressions as "human avalanche" or "flood of asylum seekers." Of course, when compared with the neighbouring countries of the real crisis flashpoints and war areas in the Middle East, from which most of the people escaped to Europe in the 2010s, the term "crisis" was a gross exaggeration, and yet attitudes were hardened. On the one hand, there were those Finns who sought to help the migrants. They volunteered for the Red Cross and other non-governmental organizations (NGOs) and set up informal aid groups. They also gave donations for food and clothes. On the other hand, the political attitudes became unwelcoming and the right-wing populist Finns Party gained support among the electorate and entered the government in 2015. Furthermore, extremist organizations attempted to capitalize on the situation, patrolling the streets of various towns and cities where the migrants were placed to keep the Finns "safe" from asylum seekers. Finnish social media was abuzz with racist reactions. There was also a certain amount of fear, especially among young women who worried for their safety, after hearing rumours of assaults and rape in the Finnish press and social media. Against this background, the country became somewhat polarized in attitude and understanding.

New reception centres were quickly set up to house the migrants. Remembering those moments in Finland, we can recall the reactions of some of the people living close to the centres: both welcoming and hardened atmospheres co-existed. The situation was new for many Finns, because such migration was quite unheard of. Many migrants who had come to Finland in the past came through either a more organized process or were so small in numbers that the Finns' reaction was limited. Thus, the "crisis" of 2015 sent a shock wave through the life pulse of Finnish society, a country with a tenuous immigration history. However, after the year 2015, the situation returned to normal in terms of applications for international protection; their number varies from 3000 to 5000 per year.[2]

Reception centres are the primary places in which asylum seekers live in Finland. Centres are located across the country, from Helsinki, the capital of Finland, on the south coast, to the city of Rovaniemi in the north. There were a total of 38 reception units in operation for 7000 people in September 2019. Besides the centres, asylum seekers live in private accommodation, especially in the Helsinki area. In total, almost 9000 adults or family members were part of the reception system in Finland in September 2019.

Mixed methods approach

The empirical data were collected in an ethnographic manner, with the concept of "thick description" (see Ponterotto, 2006) in mind, through participant observation and eight theme-centred interviews in one of the reception centres in south-western Finland in 2010. We wanted to know about the lives of asylum seekers not only in their own voice, but also from their own point of view in the particular context of a reception centre. The data were collected before the year 2015 and can perhaps be considered outdated. However, as we compare our findings with more recent studies (e.g. Onodera, 2017; Haverinen, 2018; Petäjäniemi, Lanas, & Kaukko, 2018; Nykänen et al., 2019), it is relatively safe to assume that the general conditions, and residents' worries, in the reception centres have remained the same throughout the 2010s – only the numbers of asylum seekers and reception centres have varied over the years.

The interviews were preceded by month-long fieldwork at the reception centre in the summer of 2010. The interviews with residents were conducted in August–September 2010. The fieldwork and the interviews were conducted by research assistant Jaana Salo. The field notes were used only as background material for the analysis. The interviews contained five themes: everyday life in the reception centre, residents' participation and hobbies, reception centre activities and human relations, life management, and knowledge about Finnish society. The employees of the reception centre were also interviewed, but those interviews are not included in this chapter. The first author of this chapter decided the interview themes, conducted one of the interviews, visited the reception centre over a 1-month period in order to familiarize himself with the research location, and discussed the progress of the interviews with the research assistant. Thus, it was possible to negotiate the meanings of the interviews and increase shared understanding and the reliability of the analysis.

It is obvious that the situation of asylum seekers who lived in the reception centre without a decision on international protection was not ideal for the purpose of our research. It was natural that the interviewees were hesitant to bring up issues that, in their minds, could somehow hamper their possibility of receiving international protection. They were told that the interviews had nothing to do with the Finnish Immigration Service, the police, or other authorities – and that they could freely express their ideas. It was also guaranteed that their names would not be mentioned in the text. At their request, the interviews were not recorded. All of the interviews but one (which was done in English) were conducted in the native language of the interviewees, and an interpreter was used. The interpreter might have raised some suspicions, too, as the residents might think that he represented "the system" and it would not be in their best interests to talk openly. In addition, it might have been difficult for women to express their opinions in the presence of a man of their own nationality. Our impression was that, in these conditions, the interviewees expressed their thoughts freely and used this opportunity to open their hearts.

We sought to interview individuals with different personal and asylum histories to highlight the voices of residents in different situations. As a result, the interviews included both those who had been living in Finland for more than 2 years and those who had lived in the reception centre for a few months. However, we were not interested in their age or gender identity and, therefore, did not want to register them. The choice was "political": leaving such classifications aside increased the universality of the interviews; each interviewee spoke for themselves, for everyone else, and anyone. Thus, they represented all asylum seekers and, in principle, they were "anyone."

We also followed this ontological premise in the original analysis of the interviews and applied holistic content analysis (Lieblich, Tuval-Mashiach, & Zilber, 1998). In brief, the process of analysing the interviews in a holistic manner consists of reading them several times "until a pattern emerges, usually in the form of foci of the entire story." As Lieblich, Tuval-Mashiach, and Zilber (1998) state, "a special focus is frequently distinguished by the space devoted to the theme in the text, its repetitive nature, and the number of details the teller provides about it" (pp. 62–63). In this text, the focus is on "experiential entity" – that is, on the search for the *leitmotif*, the leading or guiding message of the interviews. What, then, was the leitmotif of the interviews?

Leitmotif: living in nowhere

People living at the reception centres around Finland have suffered from many ills before their arrival in Finland: they have lost their family members, experienced religious persecution, and witnessed the horrors of war. Thus, they often suffer from psychosocial traumas and are in a vulnerable state. The migrants have come to Finland by land, by sea, or by air, and alone or with their partners – perhaps with their families. First, they seek the border control or police and declare that they need asylum or international protection. Second, the authorities register them as asylum seekers by entering their basic details (name, date of birth, country of birth) and by taking their fingerprints, signature, and photograph in the information register. After that, the migrants are directed to the reception centre. There, the waiting begins. Later, they meet with an immigration officer in an asylum interview with the Finnish Immigration Service. Some of them receive deportation notices immediately, without any consideration of asylum. It needs to be said that Finland expels most asylum seekers.

At the reception centre, people are not living on top of a hill where one can see the valley below; rather, they are living on top of nothing but the abyss. It seems as if the asylum seeker has arrived from nowhere and is nowhere. Usually she is waiting, for months, sometimes for years, without work or anything worth doing. We often learn very little about her dangerous journey, or her life before it, or the hardships she experienced. The newcomer does not speak Finnish. Her ultimate goal is to survive today, then the next day, and the day after that. And so it goes, an endless journey into the abyss. Her devoted wish for asylum or any positive residence permit follows her in her dreams if she happens to get some rest from her many worries. She hopes for her own home, work, livelihood, and ordinary life, just like the rest of us. But what she gets instead is a feeling of nausea and eventually slides into the state of anomie or stasis, of living in limbo, or of "suspension of time" (Griffiths, Rogers, & Anderson, 2013, p. 19) while the world around her continues forward. However, the reception centre is usually a better place than the country the residents were forced to leave in terms of risk of violent death. In the centre, there is no such risk, and, in general, the residents feel secure at the centres in Finland (see Koistinen, 2015, p. 60).

> Life at the reception centre is safe. There's tight space here. I've been here for almost two years, but I have not yet received a decision in my asylum case. There are a lot of people here and we have security and food. At two, three o'clock at night I usually only get to sleep, but the kids go to bed early, around nine o'clock. I wake up in the morning at seven or half past eight, send older kids to school and cook for them in the morning and back at school, three meals a day. I have young children and the day usually goes to feed them, or we watch TV. Sometimes I have a good day's sleep.
>
> (Fadi; all the names of the interviewees, from here on,
> are pseudonyms to avoid identification)

But the centre lacks everything that is known and familiar to its residents. Depression and anxiety are common as the seekers wait for a decision on their asylum claims. They often feel they are living in *nowhere or in a great wasteland*. The reception centre is a kind of *Terra nullius*, nobody's land. It is an area that nobody dwells in, at least, no one can appreciate. The residents are no longer in their homeland, but also in no part of Finnish society. They live in a state of transition in time: no longer in their home country (or where their journey began), but not yet in a new country. An everyday difficulty is the lack of a common language: it portrays the general challenges of a multicultural world, which of course are not historically insurmountable. As one of the residents, among many others, states: "My neighbors in this same unit come from different countries. As I only speak my mother tongue I cannot to discuss with them, because we have no common language" (Sabeir).

In these respects, we may need to admit, as Italian philosopher Giorgio Agamben has put it, that a reception centre resembles a camp:

> If this is true, if the essence of the camp consists in the materialization of the state of exception and in the subsequent creation of a space in which bare life and the juridical rule enter into a threshold of indistinction, then we must admit that we find ourselves virtually in the presence of a camp every time such a structure is created, independent of the kinds of crime that are committed there and whatever its denomination and specific topography.
>
> (Agamben, 1998, 98)

Like life inside a camp or a prison, the reception centre is a control system. The centre's control system includes the locking of the house entrance and also the common area doors. The supervisors, who have keys to the locks, hang them around their neck as symbols of their power. In this respect, the place resembles an open prison. The difference is that time spent in prison usually has an amount of time that can be predicted based on a judge's judgment. The rush of despair, anticipation, and the deserted atmosphere blur the life of the reception centre. It is interrupted by intermittent bursts of words and mêlées that require guidance, sometimes by the police. Because of the control system, the inhabitants are virtually devoid of any status as social and political beings, and this has its side effects.

> I try to live in peace and get along with everyone, but there are many people from different countries and cultures and they live in the same place. [...] Unfortunately, if somebody has lived here for a long time and becomes the boss of that department, then the regulations of the staff will not go through.
>
> (Aazar)

The oppressed people (who are fleeing their home countries) who are living in the reception centres are often the main sufferers, as predatory capitalists and warmongers teach us that the world is entering an era of permanent crisis – or, rather, an "economic emergency" – with new reasons to cut benefits to pensions, health services, and education. At the same time, the political horns of the nationalist extreme right reinforce their populist hues by making asylum seekers and other distressed scapegoats their prime enemy. While anti–immigration populism has gained popularity (the Finns Party is at the top of the polls, with 20 per cent in September 2019) in much Western party politics and civic sentiment, calling for tougher rules, enhanced border controls, arrest by security guards, and the expulsion of migrants labelled illegal, a longer-term development is also evident. Over the last 30 years, relatively open reception and permanent repatriation of migrants from the East to the West have been abandoned and replaced by an exclusionary asylum policy of keeping 'strangers' away from Finland.

In spite of the "structural forces" that prevail in the centre, the rules imposed from above, and the outward-looking exterior of the closed doors, the residents nevertheless seek to manage their lives. One way – maybe a way not to actually manage life, but to cope with it – is to resort to passivity, another way is stealing, and the third way is using antidepressants – measures that we would not define as means of life management, but find bizarre and even insane. But just when our reasoning is that, "they must be a little crazy or they are stupid," we need to realize that we may not know enough about them and their lives. Then, as Howard S. Becker (1998, p. 28) reminds us, "[i]t's better to assume that it makes some kind of sense and look for the sense it makes." At first glance, we may think that such destructive behaviour is caused by a psychological disorder. But, if the context of the asylum seekers' life and their conditions in the centre are taken into account, then we must come to a conclusion that their seemingly odd behaviour is a symptom of almost total social exclusion and social exhaustion. Therefore, as Pihlaja states, "integration matters should begin right at the start of the reception process" (Pihlaja, 2017, p. 41). We contend that the humanity of the asylum seeker or migrant must be recognized from the very beginning of the process. Integration is one necessary ingredient, but we also need to openly discuss what further ingredients would make reception centres more humane.

Although the overall picture our interviewees paint is rather gloomy, there were signs of being-in-the-world and active agency. A father of a 4-year-old daughter was a dressmaker and had sewn dresses for his daughter. Another resident was a proficient fishman who went fishing in the nearby lake. These sort of activities were not part of the centres' official schedule, but the residents' informal daily rhythm that helped them to organize their life and gave them, in their familiarity, a sense of belonging, perhaps a certain sense of home, too. In this respect, as Rebecca Rotter (2016) has pointed out, waiting for a resolution of their immigration status, although often alienating, can also be active as it structures residents' time with a variety of routines and self-directed projects.

Despite the social exclusion and living in "nowhere," the interviews also conveyed a different story, a brighter sense of being: residents often told, through opposites ("what's wrong and worried"), how they are active in dire conditions, and not only react to the poor environment, but also believe and hope. And the mere gesture of telling, the plain decision to speak or not to speak in the interview, contains a sense of freedom, an important element of a worthy life. Other elements of a worthy life include self-realization and opportunities for education and culture.

> I like music and dancing. When I am in a good mood I dance in my room. I would like to go to dance class to dance Latin dances. And I like to play tennis, but it is not possible here.
>
> (Nerida)

In addition a worthy life includes pondering one's identity ("who I am, where I am"); being together, loving (including reproduction); using small opportunities to work to earn a living, political participation, and community influence; and "biographical and narrative work" of one's own life – that is, the opportunity to tell others about one's life history and aspirations (see Sennett, 2003, p. 243). Through the moments of pain and anxiety, there are moments of wellness and well-being too, and, yet, there are moments of living reality that linger in the memory of the past and also now in the present.

> I would hope that we could get a residence permit and live safe life in calm and peace. I would like to attend school and afterwards get a job. Before coming to Finland I studied in high school and we had a chance to study biology, chemistry, physics and laboratory sciences. Maybe I would like to study medicine.
>
> (Elaha)

Despite the past experiences and future hopes, an unfortunate result remains: the current reception system transforms asylum seekers into passive recipients of various ready-made procedures. The system seems to treat these people as unskilled and dependent individuals, although many of them are professionals in different fields. Halleh Ghorashi (2005, p. 195) has put it aptly:

> In this way, refugees waste potentially the most effective years of their lives in a new country in isolation and passivity. An active life in the early years of their exile could help them to distance themselves from the past and to put energy into building a new life in the new country. However, an isolated form of reception not only destroys these years, it also contributes to the situation in which refugees can become prisoners of the past. As a result, people whose condition of survival is to be active and productive are reduced to people who are passive and so become burdens on society.

What is to be done in reception centres?

Waiting in the reception centre need not be passive and lead to an anomie-like existence. It can also be a real human experience, where the human being can actually feel and act fully in the world and not live in uncertainty, where everything is unchangeable and controlled and fixed in advance. A "being-in-the-world experience" has many consistencies. One of them is the empathic impulse, the ability to place ourselves in another's shoes, that should be at the core of our life-world, especially in regards to reception centres. The implication of this impulse is the sense of "aliveness," one of "caring for, knowing, responding, affirming, enjoying" our lives (Fromm, 2008, p. 37). The newcomers are certainly beings in the world and not commodities for objectification. Thus, the reception centre must be a decommodifying aspect of living for all present, whether migrants or workers and volunteers. Reception centres need to be places of building an authentic agency that can "provide the advantage of reflection and feed the power of imagination" (Ghorashi, de Boer, & ten Holder, 2018, p. 387) – in a word, the centre ought to provide the residents ingredients for a new beginning in new living conditions. The following list of recommendations, based on our empirical findings, is part of a solution to the alienation and marginalization that exist in reception centres and in the asylum process in general.

Supplies for self-study and action

Many of the residents' aspirations for educational and cultural opportunities can be realized in very simple moments of being. Bicycles are needed so that they can learn how to ride a bike; sewing tools are needed for crafts; self-study of language, culture, and society, in turn, requires textbooks and a functioning internet connection. Omar (pseudonym), an asylum seeker who has been living at the centre for 5 months, presents very thoughtful ideas about studying Finnish. The problem for him and other residents, who only speak their mother tongue, is that they do not get the most out of the language teaching given in Finnish in a group that does not have other students speaking the same language. Finnish language studies require active peer support, so that it would be possible to ask a friend when needed. In addition, Omar makes a sensible suggestion: at the reception centre, Finnish textbooks and proper dictionaries could be distributed as self-study material. The idea corresponds to a familiar, recognized practice in development cooperation: give people tools, not money.

Even with small material inputs, big improvements can be achieved. Some residents of the reception centre would like to teach others their skills. This would require mapping residents' knowledge and skills in the initial interview and then, accordingly, taking action to make such teaching possible. There are also many skilled people among the asylum seekers who are ready to go straight to work to contribute to the greater community. Personal recognition, recognition of their skills, and especially designed "bridge studies," complemented by, for example, cooperative and other entrepreneurial training, could help refugees and migrants to get started in the Finnish labor market.

More information about Finland

The residents wanted information about Finnish society, which they did not get at the reception centre. According to the employees, things are solved one by one, when problems arise. However, you should be able to anticipate situations to avoid problems. This is why they need training that enables them to act independently and thoughtfully in different situations. In addition, a good quality internet connection allows residents to stay in touch with their relatives and friends and to create online communities where they can share information and help each other cope with living in foreign surroundings.

> I haven't learned much from Finland here. It is understandable as we have lived all the time inside the reception centre and we have not yet entered society. I think that we will learn from the Finns a lot of things that we have not yet learned, but it will happen only when we have moved away from the centre. There is nothing but fighting and sorrow here. I would have liked to have a course about Finnish society, but that has not happened. At another reception centre, a Finnish teacher told me something about Finnish history and more, but that is all I have been told.
>
> (Kaaseb)

Communal dining and community farming

Cooking and eating are among the everyday ways of trying to manage life. They can increase life control, but, on the other hand, they can also isolate residents in separate families if they cook in their own rooms. In some centres, meals are prepared for the residents, and families do not prepare their own food. This could add to the exclusion and marginalization that people might feel, as their own experience is not valued. There are benefits to eating together, especially if the residents are allowed to prepare their meal together for everyone, in turn. Many reception centres, especially in rural areas, have the potential for home-grown farming, and perhaps also for community farming, which would enable residents to interact with other people in the area and earn some money in a society where work is valued.

Conflict tolerance and resolution skills

Social psychological difficulties occur within the family ("living in a reception centre is so difficult that it has really badly affected our relationship") and outside (in disputes about cleaning or using the toilet and shower). Although the reception centre is neither an organized nor a functional community, group dynamics still exist. There are opinion leaders and also people who know how to deal with employees by pulling on the right ropes. In conflict situations, it may be easier for the employees to wait until the situation calms down than to uphold justice and agreed rules. The employees of the centre are required to develop conflict management and resilient skills. And, in general, the social ethos of the centre would require the development of mutual respect and the "generalization of interdependence" (see de Swann, 1988; Sennett, 2003).

Facing desires and reality

Some residents have clearly unrealistic views of Finland and what they know and can do in Finland. They are unfamiliar with the Finnish society, what kind of knowledge and skills are required in order to study in school and cope in an advanced information society like Finland. If basic education and language skills are lacking, it is difficult to meet the expectations of a highly skilled work life or university studies at the early stages of official integration process.

Community and its rules

Many family members were concerned about their children's upbringing in an unknown society. There were also situations where parents, owing to lack of language skills, did not have the opportunity to address their children's problems – for example, at school. They lack "community ownership," the ability to take care of things in common – in this case, what children can do and how they are treated. The same lack of coordinated and negotiated codes of conduct is also reflected in how cleaning becomes a life-changing topic in interviews, in the reception centre's everyday life, and in resident meetings.

Reception as mutuality and participation

Many of the asylum seekers' educational and cultural needs can be materially addressed, but there is also a need for a participatory culture where everyone is equal when they come to Finland. Creating such a culture is an educational process that can be mobilized through methods that engender and empower the inhabitants (and workers) in ways similar to cultural education and sociocultural animation. One such process we believe is the importance of mutual dialogue and mutual experience, especially when we consider the concept of "sympathetic understanding." According to Patricia Shields, a sympathetic understanding "helps make sense of the experiences of others and thus facilitates meaningful communication and social change" (Shields, 2006, p. 431). The goal in terms of a "social claim" would be to engage each other to achieve the dignity of everyday life through the prism of human solidarity and cooperation (Shields, 2006). The educational need for both the Finns and migrants would be based on the desire to share lived experiences for a better understanding of the other. In this respect, we would all benefit from applying Paulo Freire's ideas and his dialogical methodology in our social practices to gain a critical consciousness, where we can see not only some of the oppressive elements of our own life-world, but also our commonality as human beings (Freire, 2005; see also Suoranta, 2019).

Reception centre as a community centre and a village square

The psychological difficulties of the inhabitants have often begun in their countries of origin and exile. In Finland, the lack of sufficient psychological services for asylum seekers does not improve the situation. However, forms of help such as therapy are not the only means. There are also various functional methods, mental activation, future work, and drama exercises to work on and "outsource" problems. Reception centres should be invested with the power of the municipality and the organizations involved in the overall development of the subregion.

Reception centres could also be a village square for new Finns or citizens of the world, combining diverse forms of cultural activities. In this scenario, the social and cultural needs of families could be taken care of, as well as childcare, which often suffers from a lack of resources. A village square could exploit the potential of residents and support their collectively organized childcare, and the parents would all be able to take a moment's respite. Just like any village square, the reception centre could be a vibrant community of people who come together in conversation and in activities. They would share their small and large moments of existence with a democratic spirit within a diversified human existence. In this, it becomes important to also get the migrants' point of view about their experiences in their new country (Pihlaja, 2017). The village square can be a meeting point of deliberative engagement for mutual learning through shared experiences of empathy.

Conclusion

In this chapter we have demonstrated, based on empirical data analysis, that the interviewed asylum seekers experienced their life in the Finnish reception centre as passive waiting in an anomie-like situation, and that living in these conditions resembled life in a camp. This is partly because of Finland's rather restrictive reception system, accompanied by oft-racist images and attitudes towards "strangers." On a more general level, we would claim that how asylum seekers are treated in the reception centre has its roots in the lack of social and political imagination to go beyond the usual in how to deal with asylum seekers. It is as if the staff of the reception centre are used to not having the ability to act unless there are clear administrative guidelines for action: it is far easier, and perhaps safer, to say "no, it is not possible," "don't do," or to remain altogether passive than to garner the courage to act by using the thinking and action tools and ideals necessary for positive social change. The basic idea is very simple: when the concrete situation is at hand, the idea of equality or justice requires the immediate realization of the ideal, without any delay in legal, administrative, or public justification or implementation, and before the action based on the ideal is considered publicly acceptable and feasible. "It is the deliberate striving towards realization itself that will convert the impossible into the possible, and explode the parameters of the feasible," as Peter Hallward (2010, p. 112) has stated.

We cannot expect the Finnish nation state to change its rather restrictive reception system soon, or that the capitalist world order that reproduces the oppression and wage and class wars and that forces people to leave their homelands would be abolished in short order. Instead, what we can expect, and what we need to do, is to take step-by-step approaches and actions in a street-level civil society – that is, with various NGOs and grassroots movements – to come up with and explore new policies and practices – with or without the help of the welfare state. These initiatives could give rise to autonomous regions and practices within the system that may in time help to transform the current social and political system that so obviously produces unintelligence, intentional or unintentional misunderstanding between people, and right-wing populist-nationalist leaders who use nation-states to wreak global havoc. Perhaps, indeed,

> the time has come to start creating what one is tempted to call liberated terri-tories, the well-defined and delineated social spaces in which the reign of the System is suspended: a religious or artistic community, a political organiza-tion, and other forms of "a place of one's own."
>
> (Žižek, 2008, p. 426)

Non-governmental and grassroots civil society has a lot to do with asylum seekers and other people in need, and it should be guided to create a living space in which people and planet come before profit. Civil society

> enables people to engage in activities of their own interests and to focus on issues which are important for them both personally and communally. People also help and support each other through the operations of civil society, and in doing so, they create social security.
>
> (Harju, 2006, p. 71)

That is, even if we knew that the idea of a state also had to be changed to corre-spond to a "changing world," we can still keep our hope alive for a better future, and a more socially oriented habitat, within the living and convivial spaces of civil society. NGOs and grassroots movements can often offer the needed "flexibility, adaptability and collaborative ability" allowing them "to connect and employ their diverse resources" and "build further on those collaborative structures" in helping refugees (Rast, Younes, Smets, & Ghorashi, 2019, p. 13). They can also be "open to suggestions and contributions from refugees and locals" and, by doing so, "not only provide newcomers with basic help but also make them feel appreciated and comfortable" (Rast, Younes, Smets, & Ghorashi, 2019, p. 13).

Therefore, we need civil society volunteers who are ready to create "liberated zones" that allow people to cooperate and reciprocate in safe spaces. The informa-tion, knowledge, and wisdom in these zones must be accessible directly, without bureaucratic decision-making. In volunteering, everyone should be equal: that is, people are not objects but historical subjects of their own actions. These principles have radical consequences, as French philosopher Alain Badiou states:

> A first consequence is the recognition that all belong to the same world as myself: the African worker I see in the restaurant kitchen, the Moroccan I see digging a hole in the road, the veiled woman looking after children in a park. That is where we reverse the dominant idea of the world united by objects and signs, to make a unity in terms of living, acting beings, here and now. These people, different from me in terms of language, clothes, religion, food, edu-cation, exist exactly as I do myself; since they exist like me, I can discuss with them – and, as with anyone else, we can agree and disagree about things. But on the precondition that they and I exist in the same world.
>
> (Badiou, 2008, p. 39)

But, as Badiou further points out, the time at hand is still the time of nation-states with their borders and restrictions, and there is no guarantee that the situation will change very soon. Therefore, a political process on asylum seeking and asylum seekers and all the others in their position must start with the following elementary questions:

> The question of how, concretely, we treat the people who are here; then, how we deal with those who would like to be here; and finally, what it is about the situation of their original countries, that makes them want to leave. All three questions must be addressed, but in that order. To proclaim the slogan, "an end to frontiers," defines no real policy, because no one knows exactly what it means. Whereas, by addressing the questions of how we treat the people who are here, who want to be here, or who find themselves obliged to leave their homes, we can initiate a genuine political process.
>
> (Badiou & Hallward, 1998, p. 117)

Notes

1 See the Finnish Migration Service statistics: https://tilastot.migri.fi/index.html# applications/23330?l=en&start=540&end=550
2 SSee the Finnish Migration Service statistics: https://tilastot.migri.fi/index.html# applications/23330?l=en&start=540&end=550

References

Agamben, G. (1998). *Homo Sacer*. Stanford, CA: Stanford University Press.
Badiou, A. (2008). Communist hypothesis. *New Left Review*, 49, 29–42.
Badiou, A. & Hallward, P. (1998). Politics and philosophy – an interview with Alan Badiou. *Angelaki: Journal of the Theoretical Humanities*, 3(3), 113–133.
Becker, H. S. (1998). *Tricks of the Trade*. Chicago & London: University of Chicago Press.
de Swann, A. (1988). *In Care of the State. Health Care, Education and Welfare in Europe and the USA in the Modern Era*. Cambridge: Polity Press.
Freire, P. (2005). *Pedagogy of the Oppressed*. London & New York: Continuum.
Fromm, E. (2008). *To Have or to Be*. London & New York: Continuum.
Ghorashi, H. (2005). Agents of change or passive victims: the impact of welfare states (the case of the Netherlands) on refugees. *Journal of Refugee Studies*, 18(2), 181–198.
Ghorashi, H., de Boer, M., & ten Holder, F. (2018). Unexpected agency on the threshold: asylum seekers narrating from an asylum seeker centre. *Current Sociology*, 66(3), 373–391.
Griffiths, M., Rogers, A., & Anderson, B. (2013). *Migration, Time and Temporalities: Review and Prospect*. COMPAS Research Resources Paper. Retrieved from www.compas.ox.ac.uk/wp-content/uploads/RR-2013-Migration_Time_Temporalities.pdf
Hallward, P. (2010). Communism of the Intellect, Communism of the Will. In Douzinas, C. & Zizek, S. (Eds.) *The Idea of Communism* (pp. 111–130). London: Verso.
Harju, A. (2006). *Finnish Civil Society*. Helsinki: KVS Foundation.
Haverinen, V.-S. (2018). Pakotetusta toimettomuudesta – turvapaikanhakijoiden kotoutuminen vastaanottotoiminnan osana. *Janus*, 26(4), 309–325. doi:10.30668/janus.76386

Koistinen, L. (2015). Vastaanottokeskukset, onnistuneet käytännöt ja turvapaikanhakijoiden kokemukset. In Jauhiainen, J. (Ed.) *Turvapaikka Suomesta? Vuoden 2015 turvapaikanhakijat ja turvapaikkaprosessit Suomessa* (pp. 49–64). Turun yliopisto, Turku: Turun yliopiston maantieteen ja geologian laitoksen julkaisuja 5. Retrieved from http://urmi.fi/wp-content/uploads/2017/04/URMI-1.pdf

Lieblich, A., Tuval-Mashiach, R. & Zilber, T. (1998). *Narrative Research. Reading, Analysis, and Interpretation*. Thousand Oaks, CA: Sage.

Nykänen, T., Koikkalainen, S., Seppälä, T., Mikkonen, E., & Rainio, M. (2019). Poikkeusajan tilat: vastaanottokeskuset pohjoisessa Suomessa. In Lyytinen, E. (Ed.) *Turvapaikanhaku ja Pakolaisuus Suomessa* (pp. 161–182). Turku: Siirtolaisuusinstituutin tutkimuksia 2. Retrieved from https://siirtolaisuusinstituutti.fi/wp-content/uploads/2019/12/t-02-isbn-978-952-7167-60-1-turvapaikanhaku-ja-pakolaisuus-suomessa.pdf (accessed April 3, 2020).

Onodera, H. (2017). The Diversity of Waiting in the Everyday Lives of Young Asylum Seekers. In Honkasalo, V., Maiche, K., Onodera, H., Peltola, M., & Suurpää, L. (Eds.) *Young People in Reception Centres* (pp. 75–79). Helsinki: Finnish Youth Research Society, Finnish Youth Research Network. Internet Publications 12. Retrieved from www.nuorisotutkimusseura.fi/images/nuorten_turvapaikanhakijoiden_elamaa_en_web_090118.pdf

Petäjäniemi, M., Lanas, M. & Kaukko, M. (2018). Osallisuus turvaa hakevan reunaehdoissa. Hätämajoitusyksikössä asuvien nuorten miesten kertomuksia arjesta. *Aikuiskasvatus*, 38(1), 4–17.

Pihlaja, S. (2017). Young Asylum Seekers Need Diverse Forms of Integration. In Honkasalo, V., Maiche, K., Onodera, H., Peltola, M., & Suurpää, L. (Eds.) *Young People in Reception Centres* (pp. 39–46). Helsinki: Finnish Youth Research Society, Finnish Youth Research Network. Internet Publications 12. Retrieved from www.nuorisotutkimusseura.fi/images/nuorten_turvapaikanhakijoiden_elamaa_en_web_090118.pdf

Ponterotto, J. (2006). Brief note on the origins, evolution, and meaning of the qualitative research concept thick description. *The Qualitative Report*, 11(3), 538–549.

Rast, M., Younes, Y., Smets, P., & Ghorashi, H. (2019). The resilience potential of different refugee reception approaches taken during the "refugee crisis" in Amsterdam. *Current Sociology*, 1–19. doi:10.1177/0011392119830759

Rotter, R. (2016). Waiting in the asylum determination process: just an empty interlude? *Time & Society*, 25(1), 80–101.

Sennett, R. (2003). *Respect in a World of Equality*. New York: W. W. Norton.

Shields, P. (2006). Democracy and the social feminist ethics of Jane Addams: a vision for public administration. *Administrative Theory & Practice*, 28(3), 418–442.

Suoranta, J. (2011). *Vastaanottokeskus [Reception Centre]*. Helsinki: Into Kustannus.

Suoranta, J. (2019). *Paulo Freire. Sorrettujen Pedagogi* [Paulo Freire. Pedagogue of the Oppressed]. Helsinki: Into Kustannus.

Žižek, S. (2008). *In Defense of Lost Causes*. London & New York: Verso.

Part IV

Human rights and indigenous communities in the Arctic

8 Embodying transience

Indigenous former youth in care and residential instability in Yukon, Canada

Amelia Merhar

Introduction

How is movement from the past carried on, lived, and felt in the present? What might the long-term consequences of repeated systemic displacement be, at the scale of the body, emotions, relationships, and health? How does experiencing numerous stressful moves of home and school during the transition and identity-formation period of childhood and youth fundamentally change a person?

This chapter explores these research questions from a human geography perspective, combining insights from the mobilities paradigm with emotional and children geographies. The Moving Home: The Art and Embodiment of Transience Among Youth Emerging from Canada's Child Welfare System project worked with 15 former youth in care (ages 18–30) as paid co-researchers, employing participatory, arts-based, and indigenous research methods. The cities where the two teams conducted research were Toronto, Ontario, and Whitehorse, Yukon. These sites generate an expansive geographical understanding of the Canadian child welfare system and mobilities combining urban/suburban, and northern/rural experiences. This chapter focuses on the context, findings, and policy implications of the Yukon co-researchers as best placed to speak to the many themes of Arctic immigration, including the nuances of northern resiliencies, questions of mobilities, and justice in an inclusive society.

Whether and how the child welfare system in Canada is creating more transient subjects is a social issue that warrants discussion, especially when considering cross-overs into the criminal justice and shelter systems. Moving Home has garnered seven news articles and radio interviews discussing the co-researchers and their art, and the nexus of social issues and communities that the child welfare systems in Canada impact. Although there can be many valid critiques of media (mis)representation, this significant amount of public interest illustrates that these combined methods and questioning of embodied transience in an age of increasing mobilities and precarity touch on an issue with broad social resonance. The chapter begins with an overview of the child welfare system in Canada, followed by a summary of the theoretical framework and methodological approach, then discussion on the embodiment of transience and findings, and concluding with policy recommendations for youth transition and integration in the north.

Child welfare in Canada: indigenous over-representation, outcomes, and placement bouncing

Canada has more youth in the child welfare system per capita that many other developed nations, including the U.S., Japan, France, Italy, Norway, and Australia (Thorburn, 2007), and apprehensions of youth into care have been on the rise. There are an estimated 62,428 youth in foster and group homes in Canada (Jones et al., 2015). There are more than 300 different organizations that provide child protection and family services in Canada, under the jurisdiction of individual provinces and territories (Sinha & Kozlowski, 2013). The child welfare system and its failings are especially devastating for indigenous communities, as First Nation, Inuit, and Métis children under 14 make up 52.2 per cent of all children in foster care, although only representing 7.7 per cent of all children and youth overall in Canada (Statistics Canada, 2016). The very first call to action in the Truth and Reconciliation Commission report summarizing how to begin healing damage from Canada's colonial legacy of residential schools for indigenous children is to reduce the number of indigenous children in care (2015). The Canadian Human Rights Tribunal recently decided that the Canadian government discriminated against First Nations children and families through its inequitable child welfare services (Canadian Human Rights Tribunal, 2016). In 2019, the federal government passed Bill C-92, an Act respecting First Nations, Inuit, and Métis children, youth, and families that promises to return jurisdiction over child welfare to indigenous communities; however, the bill does not guarantee any new funding to support this outcome. Critics, including noted indigenous child welfare activist Cindy Blackstock, say this means the bill is not as strong as it needs to be (Barrera, 2019).

Beyond the initial trauma and abuse leading to a child being apprehended into foster care, once within the system, school and home life are impacted greatly (Snow, 2006, 2008). Each move means a new home, different rules, other youth in care to live with, potentially hierarchical relations with biological children of the foster parents, and usually a new school and neighbourhood. In the case of group homes, youth must become accustomed to all the staff, their shifts and personalities, as well as sharing a home with five or more youth in care. Furthermore, youth in care experience a determined moment in time when they must be independent (often at 18 years old), ready or not, even in a social context when 42 per cent of youth aged 20–29 in Canada still live with their parents (Statistics Canada, 2015). As most parents and guardians do not immediately withdraw all support from their children at 18, this abrupt removal from the system is highly problematic and a unique feature of life and identity as a former youth in care. In Canada, approximately 6000 youth "age out" of the child welfare system each year (Fallis, 2012). Outcomes for former youth in care include: "low academic achievement; unemployment or underemployment; homelessness and housing insecurity; criminal justice system involvement; early parenthood; poor physical and mental health; and loneliness. Studies across decades, countries, policy approaches and research methodology show the same results" (Kovarikova, 2018: 4). A Swedish cohort study of former youth in care sadly showed that all former youth in care should be considered high risk for suicide compared with the general population (Vinnerljung et al., 2006). There are many reasons and social determinants of health that all factor into these difficult life circumstances for former youth in care. Focusing on placement instability in the child welfare system and considering its longer-term effects over the life course contribute to understanding the embodiment of transience in this project.

"Residential mobility" is the term most closely connected with the idea of moving home from the field of housing studies, and there are relevant works demonstrating the nuanced negative impacts of residential instability upon children (Jelleyman & Spencer, 2008; Scanlon & Devine, 2001), as, overall, movement is associated with poorer academic success, unless moving from a low- to high-income neighbourhood (Roy et al., 2014). Interestingly, the mere imagining of a more mobile lifestyle generates sadness and loneliness, with an increased desire to extend the social network (Oishi et al., 2013). Many studies on residential mobility are short-term, and so long-term trajectories are especially valuable to learn from. The lifespan biography work by Coulter and Van Ham (2013) suggests important findings on long-term residential mobility, including that it may take multiple moves to resolve housing disequilibrium, and that there are long-term negative psychological effects to always wanting to move.

Considering long-term residential mobility within ongoing colonization in Canada, indigenous peoples living in urban centres experience more frequent geographic mobility (Canadian Mortgage and Housing Corporation, 2002). Snyder and Wilson's research with urban indigenous peoples "demonstrate[s] mobility is an intergenerational phenomenon, influenced by colonial practices" (Snyder & Wilson, 2015: 181). This intergenerational aspect of residential mobility is of important note, as it contrasts strongly to general population-based residential mobility studies that often do not consider these data. The intergenerational effects of Canada's Indian residential schools and an increased likelihood of ending up in the child welfare system are confirmed (Barker et al., 2014). The Truth and Reconciliation report stated bluntly that Canada's child welfare system has merely continued the assimilation that the residential school system began decades earlier (2015). These historical events still shape the lives of indigenous Canadians today, and they are the precursors to the flawed child welfare system.

Youth are placed in an average of 7 homes during their time in the child welfare system (Covell, 2010), though many studies cite youth individually listing higher numbers including as many as 32 or 88 (Contenta et al., 2014; Serge et al., 2002). The beginning of a placement is the highest risk for breakdown, and, unfortunately, between a quarter and a half of children have three placements during their first year in care (Oosterman et al., 2007). Factors that are known to increase the number of foster and group home placements include: the age of the child (being older increases placements), being placed in a non-relative foster home, and the mental illness and behavioural diagnosis of the child (Koh et al., 2014; Konijn et al., 2019). "Placement bouncing" is a term created by youth in care to describe their instability. It has significant socio-spatial impacts, including uprooting relationships, redefining home, and often a new school to attend, switching mid-year (Youth in Care Canada, 2003). Placement bouncing is connected to even poorer outcomes for former youth in care, including greater life dissatis-faction, lower self-efficacy, and more criminal convictions (Dregan & Gulliford, 2012). Agency, or lack thereof, deeply relates to how one feels about moving, as mobilities scholars have discussed on the politics of mobility that shaped present movement (Cresswell, 2010). Beyond the physical movement to a new home, and the external relations that change and shift, youth who grew up in care describe three distinct ways of experiencing instability: observed instability, imposed insta-bility, and self-induced instability (Hébert et al., 2016). Observed instability is seeing others "placement bounce", imposed instability is being moved suddenly without any youth or child input, and self-induced instability is either consciously (or unconsciously) being difficult in order to force a placement change and can also include running away. The long-term effects of this instability feed into and can grow this latter form of self-induced instability and become embodied transi-ence (Merhar, 2017).

Building a theory of transience over time in the body: mobilities, children's geographies, and embodiment

Mobilities are infused in Moving Home, as it focuses on the issue of movement within the child welfare system and seeks to understand the long-term embodied realities of living in government care and experiencing repeated displacements. The mobilities turn in the social sciences (Hannam et al., 2006; Sheller & Urry, 2006, 2016) has encouraged interdisciplinary scholars to look at mobility for mobility's sake alone, instead of movement as transport, simply a way of getting to and from places. Furthermore, these movements can be more deeply understood by considering key parts of forces, speeds, rhythms, routes, and frictions (Cress-well, 2010). The political dimensions of mobility can be appreciated in many ways with respect to children in care, but most especially through the disproportion-ate number of indigenous youth in care in Canada. Nunavut, Northwest Terri-tories, and the Yukon have the highest per capita rates of indigenous youth in care (Statistics Canada, 2016). Research in Nordic countries tells a similar tale of inequity, particularly in Greenland, with regard to the health disparities of Inuit children and youth living in villages (Niclasen & Bjerregaard, 2007). Although 70 per cent of indigenous Sami people live in Norway, Statistics Norway does not report what percentage of children in care are Sami. However, they do report that immigrant children and children born to immigrant parents are significantly more likely to interact with the child welfare system than white Norwegian-born families (Dyrhaug, 2016). In this case, it seems that, for young people, migration to and mobility in Norway can incur additional and unforeseen risks of family breakdown.

Buliung et al. (2012) have called for geographers to explore youth mobility using a wide variety of methods and theoretical frameworks. By asking youth in care about their experiences and inviting them to participate actively as co-researchers using arts-based research (ABR) methods, Moving Home responds to Buliung et al.'s (2012) call. Combined with participatory action research (PAR), by sharing the research design, data output and action duties with young people involved in the project also bring excluded voices into geographical knowledge production (Cahill, 2007). Exploration of and inspiration by interesting work by non-mobilities geographers (Nash & Gorman-Murray, 2014) fuse the focus on movement with an embodied, long-term outlook. This exploration of the embodiment of transience focuses on the embodied practices and outcomes that relate to frequent moves of home as a young individual whose identity is in formation, which brings a need for discourse on childhood, youth, and identity.

Children's geographies encompass childhood and youth from infancy to age 25. Most research has been done with children and young people aged 6–14 (Valentine, 2003). In the global North, a common way of understanding this period of childhood is as a time of innocence and freedom from the obligations of adulthood, with common childhood experiences including day care, school, spatial and transport/movement restrictions imposed by adults for safety, and financial dependence on caregivers (Valentine, 2003). As a newer field of study that has had less time to cover all the permutations of youth, children's geographies has perpetuated the norm of children being raised by their parents, and the research has largely been conducted in the U.S. and U.K. With regard to the child welfare system in Canada, as De Leeuw (2009: 123) points out, the idea of childhood has been systematically and historically applied to Indigenous peoples, but

> little attention, though, has been paid to historic or social discourses that relegated groups of people to a perpetual state of truncated childhood while simultaneously removing their children in order that those children mature into adults who embodied radically different cultural traits than their ancestors.

This enacted and imposed childhood complicates a non-racialized and neutral child-subject, as often written about in the children's geographies literature. Exploratory work with former youth in care has borne out the importance of indigenous spirituality to belonging for former foster youth in Canada (Corcoran, 2012); however, coming from a psychological perspective, it has not fully addressed the spatiality, movement, and embodiment of placement bouncing.

In this section, the perspective shifts from acknowledgement that experiencing multiple placements as a young person in care is an inherently emotional and disruptive experience, to seeking to know how to approach and consider embodiment itself, the embodiment of transience, and embodiment in visual and performative autobiography in ABR analysis. The body is "a lived space where the consequences of violence are felt" (Bennett, 2005: 60). This quotation aptly describes an art performance as emotional, lived, and carrying long-term effects. The violence that is felt in the body can be an immediate reaction, or an ongoing, lived experience triggered from past trauma, or even self-inflicted in a combination of comforting but ultimately destructive coping mechanisms, including drugs and alcohol, as a way to make the body-as-home more comfortable (Robinson, 2005). The body as the most intimate and personal of spatial scales is the site of much social science and humanities research.

Feminist researchers have studied the body and embodiment for decades, often focusing on (but not limited to) exploring particular embodiments experienced by racialized bodies (Nayak, 2010; Veninga, 2009), gender (Braidotti, 1994; Colls, 2012), fatness (Colls & Evans, 2014), chronic illness (Moss, 1999), and queerness (Gorman-Murray, 2009). In this scholarship, the body is home and a meeting place, but it can also be an unpleasant, awkward, and alienating one. Feminist thought reminds us that performed critical autobiography works are not simply truth-telling, but a partial, embodied positionality that is performed with room for the audience to connect, enter, and reflect upon (Lather, 1993). Hesford (1999) stresses that the risks of embodied artistic self-disclosure are greater for marginalized bodies. The body as used in performative autobiography in Moving Home, in such media as self-portrait photography, spoken word, and music performance, illustrates the different ways the self and bodies can produce and share memory (Taylor, 2003).

Embodying transience in the Yukon

In the Yukon Territory, the Department of Family and Children's Services operates child welfare services. There are 14 First Nations in the Yukon, 11 of which are self-governing. Recently Yukon First Nations and the Yukon government have been working together more often to reduce youth in care and decrease over-representation. Yukon First Nations make up 23 per cent of the overall Yukon population, and 80 per cent of children and youth in care are indigenous to the territory (Yukon Bureau of Statistics, 2016; Yukon Child and Youth Advocate, 2019). The Yukon Moving Home co-researcher project ran in partnership with the Skookum Jim Friendship Centre, Yukon Child and Youth Advocate, and the Splintered Craft Youth Arts Centre, in the summer of 2016. For 2 weeks, five Yukon First Nations, Inuit, and Métis co-researchers met Monday to Friday for 6 hours a day in a local youth centre. A local elder came to open and close the research project with prayer and was available to youth for additional support if needed. Mornings were discussion- and workshop-oriented, followed by lunch, and afternoons were open for co-researchers to work on their individual or collaborative ABR projects. The project ended with an optional public art show as the action part of PAR, and everyone participated. All art remained the property of the co-researchers, and the artist statements were crucial in the analysis phase.

Exploring personal embodiment in (potentially) traumatic narratives is difficult emotional labour. With some care and awareness of the difficulty and tensions of the task at hand, the concept of embodiment of transience was shared with co-researchers through several open-ended prompt questions at the start of this research project. The questions were designed to facilitate co-researchers' connecting past movement in the child welfare system to current-day social and geographic mobility, with his/her/their ever-present body and feeling(s).

1   How do you think moving around so much, having multiple placements in group homes and foster homes, has affected you?
2   How do you feel about moving now in your life?

3   What sort of people is the child welfare system creating through so many placements?

4   What might you still be carrying with you from this experience and how do you carry it?

The first question broadly introduced the research topic, and the second offered a more concrete and focused example. Many, but not all, co-researchers readily self-identified with terms such as traveller, gypsy, always moving, or having wanderlust early on in the project, even going so far as to say they applied to participate in the research precisely because they have wondered about their frequent movement themselves and wished to learn more. Connecting these expressions of mobility to lived experience in the child welfare system, the third question is socially oriented and critical of child welfare practices. The third question strongly resonated with co-researchers, as their lived experience has made them "system-wise" (Liang et al., 2016), and they were quick to share stories, ideas, and examples of their thoughts on what sort of people the child welfare system creates. Not wanting to promote a narrative focused on former youth in care versus the Children's Aid Societies, the project then asked co-researchers to connect this lived experiential knowledge to their bodies for Question 4. Co-researchers were asked to consider the body as a home, as the home we always have, and what we carry with us all the time.

When asked by co-researchers for an example of embodiment of transience, I shared a realization that I have had concerning my own body. I explained that I had spent some time homeless after growing up in and out of foster care, sleeping outside every night, subject to the elements, other people, and the police. I'd tense up at night when I went to sleep. That tensing, my dog, and my knife were my protection when I slept. Fast forward to 5 years off the streets, and I was hired for my first job with full health benefits where I could begin getting registered massage therapy. Only then, in my first session, did I realize that, even years after I had been on the streets, my body was still tensing up every night. My body had learned a habit that was useful at the time of homelessness and transience, but now that habit was maladaptive – causing me ongoing physical pain and stress. Together, these four questions and my personal anecdote framed the way co-researchers were invited to consider their bodies, embodiment, and transience as they began their explorations and excavations of personal history.

Mixing media and methods

Youth in care are quite frequently the subject of research, as opposed to being the producers of research. PAR was chosen as a method that challenges the researcher–researched hierarchy, creating knowledge that values the lived realities of traditionally excluded people in society. Taking research beyond the academy (Cahill & Torre, 2007), the action component of PAR is integral to the method. Advocacy groups including Youth In Care Canada continually express that youth deserve more say in child welfare policies. By adopting PAR as a method, these desires for more say can be honoured. Moreover, current best practices in youth in care and former youth in care research suggest that cultivating belonging through peer relationships is integral to improved youth life outcomes (Snow & Mann-Feder, 2013). Through PAR, research subjects become co-researchers who work together and challenge social stigma.

Given the significant over-representation of indigenous youth in Canadian foster homes and group homes, indigenous research methods guided the research process. Indigenous research methods are made up of multiple epistemological and ontological beliefs, with a common theme of relationality throughout, drawing strongly on Wilson (2008), who emphasizes that research is ceremony, the importance of relational accountability, and that, ultimately, good research should change you as a person. Understanding accountability to youth participants shaped their role as co-researchers and supported designing a research project where youth capacities were enhanced by the very nature of the research process (through workshops, renumeration, skill-building, peer support, and public art shows).

Lastly, ABR methods were selected because, when traumatic memories are being considered, materiality has been shown to be a useful way to process issues. ABR can make meaning through narrative, poetry, performance, theatre, dance, film, visual arts, and music, and can be used at any and all stages of the research process, from idea generation, exploration, presentation, to dissertation (Leavy, 2008). Like all research endeavours, ABR seeks to develop new insight and knowledge that are relevant and applicable. What is so different about ABR is that theory and practice are more entwined; a living research praxis is created with the arts that acknowledges the body, the language of metaphor, emotions, and imaginaries through exploring pressing social research questions. Theatre of the Oppressed drama workshops (Conrad, 2010) and Photovoice (Castleden & Garvin, 2008) have been some of the most commonly used community arts practices in Canada. ABR excels at exploratory, descriptive, and process-based research topics, and, through these research methods, youth are telling their story in their own way, on their terms (Finley, 2008; Leavy, 2008).

Findings

Resilience as resistance

One of the most immediate themes emergent in the work of co-researchers was the desire to beat the odds stacked against them in life. These desires express a multitude of relationships and experiences to foster siblings, group home workers, social workers, family and self. Resilience framed as resistance was presented artistically as hip hop lyrics by one co-researcher, for example:

> if life is what you make it then I really got a chance
> that's my motivation when they tell me that I can't
> so tell me that I can't man, tell me that I can't they told me what I can't
> and look what it made me

Similar statements were made in photography works, spoken word, and silk-screen t-shirts. The immediate presence of this theme required additional readings on resilience, resistance, and vulnerability to better understand the work produced by co-researchers. Resilience comes through in the art and artist statements as resistance, as an intentional act of surviving in spite of the child welfare system. This manifestation of resilience from these young co-researchers is fundamentally different from a traditional definition. Place turns out to be integral to the knowledge produced, where the alienation produced through displacement in foster and group homes is more significant in an urban/suburban environment than in the Yukon. Thus, the north (or maybe smaller communities in general) exists as a protective factor against the alienation that the child welfare system produces through placement bouncing, as youth get to remain closer and connected to their cultural and spatial communities of childhood and adolescence.

Resilience as resistance is also seen through representations of indigenous culture and ontologies from indigenous co-researchers as an example of ongoing resistance against colonialism. This was present in both projects with indigenous co-researchers, but especially evident in Whitehorse, with the repeated use of clan animals in paintings and drawings, and the choice of materials and media including feathers, beads, sinew, and hide. Resilience as resistance is an interesting concept, a way to frame personal survival, if not success, in opposition to oppressive others. These others identified by co-researchers include child welfare agencies, specific social workers and group home staff, and unsupportive teachers. A conscious decision to seek a positive outcome not just for oneself, but as a demonstration of personal power against oppression has a relational connection to resilience, albeit one that is driven strongly by negatively predicted outcomes instead of hope for a better future.

Resilience is how subjects are to "be" in this precarious modern world (Evans & Reid, 2014). With regards to young people who have lived the trauma of broken families and ended up in the child welfare system, resilience is measured as the mainstream success of the young person such as stable housing, good employment, and academic success, despite the tremendous difficulties in their lives. There are strong critiques of this idea of resilience, tying it into the offloaded responsibility and entrepreneurialism of neo-liberalism. People who born into a precarious world have different response to precarity than those who presumed they would be protected, as Berlant explores in *Cruel Optimism* (2011). Expecting embodied resilience without support from people without significant material control over their lives (i.e., minors in care) is cruel in and of itself.

Gendered bodies and the natural world

Female-identified co-researchers were more likely to depict, mention, reference, and/or use the body in their work. Not only were female co-researchers more likely to depict the body, but, when they did, they frequently included multiple media or expressions, such as collage, a series, photos with a poem, to represent the complexity and entanglement of emotions, relations, and attachments.

All these multiplicities displayed what Mol (2002) describes as the body not as a single entity, and skin not as a container, but bodies extending and connecting to other bodies, practices, and technologies, which in turn affect other bodies. Feelings and bodies were depicted in manners that align with the current data on youth in care. This includes negative emotions, and bodies most often presented as alone (in terms of other human forms). Although negative emotions play a large role in many works, hope for the future and the desire to find a place called home are also present.

Although, often, bodies are alone in human form, they are frequently expressed together with representations of the natural world. The stars, the forest, mountains, flowers, and the galaxy all came up repeatedly among indigenous and non-indigenous co-researchers in the project. Nature and vast natural forces such as the stars, the galaxy, the moon, the planet Earth, as well as representation of the local landscapes respective to each co-researcher group in Toronto and Whitehorse, emerged as such an interesting and unexpected theme throughout this ABR project. In Whitehorse, the musician Beats Planet Kid has images of stars and a galaxy on his album cover, and Starchild Dreaming Loud used a background of stars for *Positive/Negative*; Michelle Charlie's *Inner Workings* has a background of stars and also depicts the moon. Furthermore, the only artist who chose to identify as anonymous drew a planet, mentioned the cloud, and a higher order in the Whitehorse group piece, which was the production of academic certificates, entitled *Knowledge Production*. Out of the five co-researcher certificates in *Knowledge Production*, one depicted a wolf, one the mountains, and one the planet and clouds. What is meant by the co-researchers collectively connecting themselves to such larger and temporally vast things as earth systems, the planets, the stars, and socio-natures (Swyngedouw, 1996)? The galaxy, the planet, and the stars are vast moving constants; even mountains emerge and erode over time. Are former youth in care, through the habit of embodied transience, turning to identify with the largest spatial home possible, which is also always moving?

In research with young people concerning ideas of home, and leisure activities in a globalized and technologically connected world, an idealization of nature emerges as a place for leisure that permits quiet reflection (Abbott-Chapman & Robertson, 2001). The multiple representations of the vast expanse of the universe by co-researchers suggest a recognition of something new in this idea of embodied transience – a scalar reach that extends beyond the Earth itself to its place in the universe. This identification with outer space might be about defining home as the broadest idea of home that is always there, and always moving as well. Harvey (2000) linked the flourishing of research on and interest in the body and globalization occurring simultaneously as a way to ground discourse against the forces of neo-liberal capitalism. Through a nuanced understanding of the body as a site of resistance (Butler, 2016), the body as existing in spatial and temporal relations must be grounded, even when the concern in this case is on repeated movement of the body becoming an embodied habit. I turn next to exploring the movement aspect and expressions of embodied transience, as former youth in care are moving on individual paths, yet within unique social worlds and imaginaries.

Movement

The embodiment of transience created by frequent and repeated foster and group home placements has resulted in a spectrum of movement and emotionality related to it. Movement is not neutral for former youth in care, whether they identify as travellers or utterly detest moving and strive to stay in one place or the other. The majority of co-researchers identified frequent moves an as ongoing and problematic issue in their lives that was tangled up with relationship issues, low income, the high cost of housing in the north, and low vacancy rates. In fact, 5 out 15 of co-researchers actually moved during the total 4 weeks of fieldwork. This is an example of the financial and housing precarity many former youth in care are still existing in and navigating.

The movements of former youth in care are complex, in direction, motivation, and temporality. Seeing freedom as an ongoing practice (Bogues, 2012) within the politics of mobility (Cresswell, 2010) allows for a new way of considering the frequently moving former youth in care as not merely subjects of poverty and oppression, but active agents in their increased mobilities, living with varying plateaus of self-awareness as to why the moving of home is not an emotionally neutral or circumstantial occurrence. Movement as freedom can be freedom-from, the practice of movement as freedom from the past and forgetting (including an avoidance of emotional intimacy inherently connected to stability). The second sub-theme of movement was a hopeful freedom-to, freedom as seeking a home, a better future. With the concept of movement as freedom-from, these artists are reacting against the child welfare system and difficult life circumstances, and movement takes them away, through forgetting and spatial distance and newness. With the concept of freedom-to, hope is also strongly present: movement is not only in direct response to and directionality away from the child welfare system, but rather toward a home and a hopeful life in the future, similar to the "geography cure," as outlined by Alcoholics Anonymous (Orrok, 1946), that a fresh start in a new place is all one needs.

Giving back and hopeful futures

Hope might be keeping some of these young people alive, and hope can also be caught up in an adventurous yet maladaptive embodied habit that protects young people while trapping them in the comfort of major upheaval and disruption. Throughout these varied relationships to hope and movement, co-researchers expressed a desire for a better life and a home in the future, not just for themselves but for youth in care overall. Giving back and community-building among former youth in care are present in a collaborative zine that was created, written to current youth in care to support them. Zine contributions noted self-care tips for current youth in care, and, in artist statements and performances, discussed the struggles and beauties of being late bloomers and assured others that they too will find their way. Former youth in care are enacting an ethics of care, creating the communities they need out of the ones that are very displaced and fragmented.

Scholars interested in an ethics of care can learn how communities such as this one are practising this theory, which can develop practical understanding, skills, and tools towards a grounded feminist ethics of care. Currently, in Canada, there is a groundswell of former youth in care activism, and that has resulted in numerous policy changes, including expanding support up to age 25 in some regions and developing tuition waivers for post-secondary support.

Conclusion and policy recommendations

The findings of Moving Home reaffirm the complex, layered, personal definition of home. This research develops a geographical approach to understanding young people in and from care in Canada, who have not, until now, been studied through a geographical lens. In the child welfare context, the notion of home involves multiple scales of workplace, institution, and home all at once (Dorrer et al., 2010). Young people in care are experiencing multiple transitions of home and school within the transition of youth to adult (Valentine, 2003). This research adds a temporal lens to the notion of home, and moving home, focusing on the embodied habits of transience learned through these multiple transitions. In terms of embodiment, Moving Home leads us to consider what else we can embody over time, besides transience stemming from multiple moves as a young person. What other embodied habits might individuals have picked up that have yet to be explored?

Concerning mobilities literature, this work builds on the politics of mobility (Cresswell, 2010) by exploring repeated systemic moves and their impact on individual lives. Repeated moves of home at the scale of the body deepen understanding of movement and mobilities. Young people who are experiencing multiple placements, then continuing to move when out of the system owing in part to embodying transience as a habit, ask mobilities research to consider the embodiment of movement and a longer relationship to temporality together. The co-researchers lived experience and art produced in Moving Home suggest that movement can be a habit, and, eventually, a maladaptive one at that.

The embodiment of transience is not currently considered when dealing with young people struggling with poverty, housing, or mental health issues. The focus is on immediate, primary, and more visible needs. The timelines do not match up for service provision from busy non-profit societies, or governments more concerned with liability and shrinking budgets. However, moving can be incredibly disruptive to continuity of care, emotional health, and well-being; conversely, so can staying in a place where one feels oppressed or stuck. Habits can linger long after they serve us well. By considering the embodiment of transience when working with displaced, unstable, and mobile populations, including immigrants, migrant workers, and refugees, these populations can be better served, supported, and heard.

The policy recommendations below are primarily for organizations, programmes, and services working with current and former youth in care that respond to the embodiment of transience. Because there is so much overlap between former youth in care and experiences of incarceration, street involvement, poverty, and homelessness, it is crucial to deeply understand the implications of the embodiment of transience within these groups as well. Colonial and neocolonial discourses still impact indigenous families today in Canada, but continuing to learn, unlearn, and use multiple actions from protest, policy, advocacy research, and awareness can move recommendations forward (McKenzie et al., 2016).

Supportive, flexible housing

Given that former youth in care are moving so much as an embodied habit, supportive housing policies should be more flexible in terms of entry and re-entry into them. Youth should not age out without housing arranged, given that 60 per cent end up homeless at some point (Gaetz et al., 2016). As an example of a step in the right direction, since 2016, Northwest Territories have added extended benefits for former youth in care, and, in 2018, 78 per cent of youth aged out with ongoing supports and income (Director of Child and Family Services, 2018).

National access to supports

Former youth in care are highly mobile across Canada. Currently, initiatives to support former youth in care are province- and territory-based, so that, if a youth moves, they lose all potential supports. Regionally extended supports differ widely across the country, from extended health care benefits in Ontario up to age 25, or a new tuition waiver in British Columbia up to age 27. A Yukon former youth in care living in either of these places would get no extended support from the governments of Ontario or British Columbia, and vice versa. Letting youth access former youth in care services across the country, instead of solely in the region where they were in care, can help support a healthier transition to independence. This includes health benefits, housing supports, scholarships, tuition waivers, and counselling. As young people live with their parents longer owing to school and the high cost of education, it is unethical to ask the youth with the most difficult backgrounds to be independent earlier than their peers.

Movement, body, and trauma-focused work

Promoting movement-based therapeutic activities for former and current youth in care can help connect them more deeply to their bodies and emotions. Addictions, from food to drugs, and fleeing unsafe situations are an embodied trauma response operating under the flight mode of our "fight-or-flight reflex" (Van der Kolk, 2015). Only by listening to and claiming the body can those stuck in a pattern of movement learn to move freely and with greater intention.

Peer support

Continued and expanded support should be offered for former youth in care to meet each other to share skills, supports, and build relationships and community. Being a former youth in care is an invisible social identity and it lasts a lifetime, because families are such an integral part of daily life. Projects such as the Voyageurs in Toronto, which begins working with foster children in Grades 5–7 to help them gain skills and friends and identify goals are key to changing the negative outcomes for former youth in care. The project also peer mentors youth throughout post-secondary studies, an important targeted intervention when only 5 per cent of former youth in care currently reach post-secondary education, compared with 54 per cent of the Canadian population (Statistics Canada, 2016).

Acknowledgements

Moving Home was supported by a Social Sciences and Humanities Research Council of Canada award, Northern Scientific Training Fund, an Association of Colleges and Universities for Northern Studies award, the Provincial Advocate for Children and Youth (Ontario), and the Yukon Child and Youth Advocate Office. Sketch Working Arts in Toronto and Splintered Craft Youth Arts Centre in Whitehorse were site partners.

References

Abbott-Chapman, J., & Robertson, M. (2001). Youth, leisure and home: space, place and identity. *Loisir et société/Society and Leisure*, 24(2), 485–506.

Barker, B., Kerr, T., Alfred, G. T., Fortin, M., Nguyen, P., Wood, E., & DeBeck, K. (2014). High prevalence of exposure to the child welfare system among street-involved youth in a Canadian setting: implications for policy and practice. *BMC Public Health*, 14(1), 197.

Barrera, J., & News, C. B. C. (February 28, 2019). Indigenous child welfare bill "path-breaking" on rights but funding still an issue, say child advocates, accessed via www.cbc.ca/news/indigenous/indigenous-child-welfare-bill-tabling-reaction-1.5037746

Bennett, J. (2005). *Empathic Vision: Affect, Trauma, and Contemporary Art*. Stanford University Press Stanford, CA.

Berlant, L. G. (2011). *Cruel Optimism*. Duke University Press Durham, NC.

Bogues, A. (2012). And what about the human? Freedom, human emancipation, and the radical imagination. *Boundary 2*, 39(3), 29–46.

Braidotti, R. (1994). *Nomadic Subjects: Embodiment and Sexual Difference in Contemporary Feminist Theory*. Columbia University Press New York.

Buliung, R., Sultana, S., & Faulkner, G. (2012). Guest editorial: special section on child and youth mobility – current research and nascent themes. *Journal of Transport Geography*, 20(1), 31–33.

Butler, J. (2016). Rethinking vulnerability and resistance. In J. Butler, Z. Gambetti, and L. Sabsay (Eds.), *Vulnerability in Resistance* (pp. 12–27). Durham, NC: Duke University Press.

Cahill, C. (2007). Including excluded perspectives in participatory action research. *Design Studies*, 28(3), 325–340.

Cahill, C., & Torre, M. (2007). Beyond the journal article: representations, audience, and the presentation of participatory action research. In S. Kindon, R. Pain, & M. Kesby (Eds.), *Connecting People, Participation and Place: Participatory Action Research Approaches and Methods* (pp. 196–206). London: Routledge.

Canada Mortgage and Housing Corporation. (2002). *Policy and Research Division. Effects of Urban Aboriginal Residential Mobility. Research Highlights. Socio-economic Series; 114.* Ottawa: The Division, 4 (Pam. Ca1 MH113 02H114).

Canadian Human Rights Tribunal. (2016). Canadian Human Rights Tribunal Decisions on First Nations Child Welfare and Jordan's Principle, Case Reference CHRT 1340/7008, availiable at: https://fncaringsociety.com/sites/default/files/Info%20sheet%20 Oct%2031.pdf

Castleden, H., & Garvin, T. (2008). Modifying Photovoice for community-based participatory indigenous research. *Social Science & Medicine*, 66(6), 1393–1405.

Colls, R. (2012). Feminism, bodily difference and non-representational geographies. *Transactions of the Institute of British Geographers*, 37(3), 430–445.

Colls, R., & Evans, B. (2014). Making space for fat bodies? A critical account of "the obesogenic environment". *Progress in Human Geography*, 38(6), 733–753.

Conrad, D. (2010). In search of the radical in performance. In P. Duffy & E. Vettraino (Eds.), *Youth and Theatre of the Oppressed* (pp. 125–141). New York: Palgrave Macmillan.

Contenta, S., Monstebraaten, L., & Rankin, J. (2014). Why are so many black children in foster and group homes? *Toronto Star*, 11, 1874–1930.

Corcoran, R. H. (2012). *Rethinking "Foster Child" and the Culture of Care: A Rhizomatic Inquiry into the Multiple Becomings of Foster Care Alumni* (Doctoral dissertation, University of Victoria).

Coulter, R., & Van Ham, M. (2013). Following people through time: an analysis of individual residential mobility biographies. *Housing Studies*, 28(7), 1037–1055.

Covell, K. (2010) The rights of the child part four: child protection and youth on the street, accessed via https://yorkspace.library.yorku.ca/xmlui/bitstream/ handle/10315/34471/Merhar_Amelia_2017_MA.pdf?sequence=2&isAllowed=y (September 2017).

Cresswell, T. (2010). Towards a politics of mobility. *Environment and Planning D: Society and Space*, 28(1), 17–31.

De Leeuw, S. (2009). "If anything is to be done with the Indian, we must catch him very young": colonial constructions of Aboriginal children and the geographies of Indian residential schooling in British Columbia, Canada. *Children's Geographies*, 7(2), 123–140.

Director of Child and Family Services. (2018). Annual report, Government of the Northwest Territories, accessed via www.hss.gov.nt.ca/sites/hss/files/cfs-director-report.pdf

Dorrer, N., McIntosh, I., Punch, S., & Emond, R. (2010). Children and food practices in residential care: ambivalence in the "institutional" home. *Children's Geographies*, 8(3), 247–259.

Dregan, A., & Gulliford, M. (2012). Foster care, residential care and public care placement patterns are associated with adult life trajectories: population-based cohort study. *Social Psychiatry and Psychiatric Epidemiology*, 47(9), 1517–1526.

Dyrhaug, T. (2016). *"Kvart fjerde barn i barnevernet har innvandrarbakgrunn"* [One out of four children in child welfare has an immigrant background] (in Norwegian). Statistics Norway.

Evans, B., & Reid, J. (2014). *Resilient Life: The Art of Living Dangerously*. Cambridge: Polity Press.

Fallis, J. (2012). Literature review: the needs of youth transitioning from protective care, and best practice approaches to improve outcomes, accessed via General Child and Family Services Authority website, https://docplayer.net/8853040-Literature-review-the-needs-of-youth-transitioning-from-protective-care-and-best-practice-approaches-to-improve-outcomes.html (accessed September 2019).

Finley, S. (2008). Arts-based research. In G. J. Knowles & A. L. Cole (Eds.), *Handbook of the Arts in Qualitative Research: Perspectives, Methodologies, Examples and Issues* (pp. 71–82). Thousand Oaks, CA: Sage.

Gaetz, S., O'Grady, B., Kidd, S., & Schwan, K. (2016). *Without a Home: The National Youth Homelessness Survey*. Toronto: Canadian Observatory on Homelessness Press.

Gorman-Murray, A. (2009). Intimate mobilities: emotional embodiment and queer migration. *Social & Cultural Geography*, 10(4), 441–460.

Hannam, K., Sheller, M., & Urry, J. (2006). Mobilities, immobilities and moorings. *Mobilities*, 1(1), 1–22.

Harvey, D. (2000). *Spaces of Hope*. Berkeley:University of California Press.

Hébert, S. T., Lanctôt, N., & Turcotte, M. (2016). "I didn't want to be moved there": young women remembering their perceived sense of agency in the context of placement instability. *Children and Youth Services Review*, 70, 229–237.

Hesford, W. S. (1999). *Framing Identities: Autobiography and the Politics of Pedagogy*. Minneapolis:University of Minnesota Press.

Jelleyman, T., & Spencer, N. (2008). Residential mobility in childhood and health outcomes: a systematic review. *Journal of Epidemiology & Community Health*, 62(7), 584–592.

Jones, A., Sinha, V., & Trocmé, N. (2015). Children and youth in out-of-home care in the Canadian provinces. *CWRP Information Sheet E*, 167. Montreal, QC: Centre for Research on Children and Families, McGill University.

Koh, E., Rolock, N., Cross, T. P., & Eblen-Manning, J. (2014). What explains instability in foster care? Comparison of a matched sample of children with stable and unstable placements. *Children and Youth Services Review*, 37, 36–45.

Konijn, C., Admiraal, S., Baart, J., van Rooij, F., Stams, G. J., Colonnesi, C., ... Assink, M. (2019). Foster care placement instability: A meta-analytic review. *Children and Youth Services Review*, 96, 483–499.

Kovarikova, J. (2018). *Review of Policy and Practice for Youth Leaving Care*. White Paper, Children's Aid Foundation of Canada.

Lather, P. (1993). Fertile obsession: validity after poststructuralism. *The Sociological Quarterly*, 34(4), 673–693.

Leavy, P. (2008). *Method Meets Art: Art-based Research Practice*. New York: Guilford Press.

Liang, B., Boileau, K., Vachon, W., & Wilton, F. (2016). *Acting in Support of Youth Voice: Theatre as Equitable Education, Relational Child & Youth Care Practice*, 28(3), 67–81. Writing with Mark: A cross-border conversation about Mark Krueger and his, 67.

McKenzie, H. A., Varcoe, C., Browne, A. J., & Day, L. (2016). Disrupting the continuities among residential schools, the sixties scoop, and child welfare: an analysis of colonial and neocolonial discourses. *The International Indigenous Policy Journal*, 7(2), 4.

Merhar, A. (2017). *Moving Home: The Art and Embodiment of Transience Among Youth Emerging from Canada's Child Welfare System* (master's thesis, York University).

Mol, A. (2002). *The Body Multiple: Ontology in Medical Practice*. Durham, NC: Duke University Press.

Moss, P. (1999). *Autobiographical Notes on Chronic Illness. Mind and Body Spaces: Geographies of Illness, Impairment and Disability*. London: Routledge.

Nash, C. J., & Gorman-Murray, A. (2014). LGBT neighbourhoods and "new mobilities": towards understanding transformations in sexual and gendered urban landscapes. *International Journal of Urban and Regional Research*, 38(3), 756–772.

Nayak, A. (2010). Race, affect, and emotion: young people, racism, and graffiti in the postcolonial English suburbs. *Environment and Planning A*, 42(10), 2370–2392.

Niclasen, B. V., & Bjerregaard, P. (2007). Child health in Greenland. *Scandinavian Journal of Public Health*, 35(3), 313–322.

Oishi, S., Kesebir, S., Miao, F. F., Talhelm, T., Endo, Y., Uchida, Y., ... Norasakkunkit, V. (2013). Residential mobility increases motivation to expand social network: but why? *Journal of Experimental Social Psychology*, 49(2), 217–223.

Oosterman, M., Schuengel, C., Slot, N. W., Bullens, R. A., & Doreleijers, T. A. (2007). Disruptions in foster care: a review and meta-analysis. *Children and Youth Services Review*, 29(1), 53–76.

Orrok, B. G. (1989). *Alcoholics Anonymous: The Story of How Many Thousands of Men and Women Have Recovered From Alcoholism. JAMA*, 261(22), 3315–3316. doi:10.1001/jama.1989.03420220129046

Robinson, C. (2005). Grieving home. *Social & Cultural Geography*, 6(1), 47–60.

Roy, A. L., McCoy, D. C., & Raver, C. C. (2014). Instability versus quality: residential mobility, neighborhood poverty, and children's self-regulation. *Developmental Psychology*, 50(7), 1891.

Scanlon, E., & Devine, K. (2001). Residential mobility and youth well-being: research, policy, and practice issues. *Journal of Society & Social Welfare*, 28, 119.

Serge, L., Eberle, M., Goldberg, M., Sullivan, S., & Dudding, P. (2002). *Pilot Study: The Child Welfare System and Homelessness among Canadian Youth*. Ottawa, ON: The Child Welfare League of Canada.

Sheller, M., & Urry, J. (2006). The new mobilities paradigm. *Environment and Planning A*, 38(2), 207–226.

Sheller, M., & Urry, J. (2016). Mobilizing the new mobilities paradigm. *Applied Mobilities*, 1(1), 10–25.

Sinha, V., & Kozlowski, A. (2013). The structure of Aboriginal child welfare in Canada. *International Indigenous Policy Journal*, 4(2) 1–21.

Snow, K. (2008). Disposable lives. *Children and Youth Services Review*, 30(11), 1289–1298.

Snow, K. (2006). Vulnerable citizens: The oppression of children in care. *Journal of Child and Youth Care Work*, 21, 94–113.

Snow, K., & Mann-Feder, V. (2013). Peer-centered practice: a theoretical framework for intervention with young people in and from care. *Child Welfare*, 92(4), 75.

Snyder, M., & Wilson, K. (2015). "Too much moving … there's always a reason": understanding urban Aboriginal peoples' experiences of mobility and its impact on holistic health. *Health & Place*, 34, 181–189.

Statistics Canada. (2015, December 22). Census in brief series: living arrangements of young adults aged 20 to 29 (Catalogue no. 98312X2011003), accessed via www12.statcan.gc.ca/census-recensement/2011/as-sa/98-312-x/98-312-x2011003_3-eng.pdf

Statistics Canada. (2016). Care stats from census 2016 www12.statcan.gc.ca/census-recensement/2016/dp-pd/prof/details/Page.cfm?Lang=E&Geo1=PR&Code1=01&Geo2=&Code2=&Data=Count&SearchText=Canada&SearchType=Begins&SearchPR=01&B1=All&GeoLevel=PR&GeoCode=01 (accessed: September 2017).

Swyngedouw, E. (1996). The city as a hybrid: on nature, society and cyborg urbanization. *Capitalism Nature Socialism*, 7(2), 65–80.

Taylor, D. (2003). *The Archive and the Repertoire: Performing Cultural Memory in the Americas*. Durham, NC: Duke University Press.

Thorburn, J. (2007). *Globalisation and Child Welfare: Some Lessons from a Cross-national Study of Children in Out-of-home Care. Social Work Monographs*. (Vol. 228). Norwich, UK: School of Social Work and Psychosocial Studies, University of East Anglia.

Valentine, G. (2003). Boundary crossings: transitions from childhood to adulthood. *Children's Geographies*, 1(1), 37–52.

Van der Kolk, B. A. (2015). *The Body Keeps the Score: Brain, Mind, and Body in the Healing of Trauma*. New York: Penguin.

Veninga, C. (2009). Fitting in: the embodied politics of race in Seattle's desegregated schools. *Social & Cultural Geography*, 10(2), 107–129.

Vinnerljung, B., Hjern, A., & Lindblad, F. (2006). Suicide attempts and severe psychiatric morbidity among former child welfare clients – a national cohort study. *Journal of Child Psychology and Psychiatry*, 47(7), 723–733.

Wilson, S. (2008). *Research Is Ceremony: Indigenous Research Methods*. Winnipeg: Fernwood.

Youth in Care Canada, (formerly National Youth in Care Network). (2003). Primer fact sheets, accessed via http://youthincare.ca/systemcapacitycont (accessed September 2017).

Yukon Bureau of Statistics. (2016). Aboriginal peoples census 2016, accessed via www.eco. gov.yk.ca/stats/pdf/Aboriginal_2016.pdf

Yukon Child and Youth Advocate. (2019). Empty spaces – caring connections: The experiences of children and youth in Yukon group care. *Systemic Review*. Whitehorse, YT: Yukon Child and Youth Advocate Office.

9 Cold temperature health risks and human rights

Stefan Kirchner and Susanna Pääkkölä

Introduction

In 2015 and 2016, the states of the continental European Arctic, Russia, Norway, Sweden and Finland, were confronted with tens of thousands of displaced persons who travelled through the region.[1] Although the number of newcomers arriving in the northernmost parts of Sweden, Norway and Finland, as well as the north-western corner of Russia, has decreased since then, the situation has raised legal questions. Far from locations commonly associated with the 2015 refugee crisis, such as Greece and Italy, the states in the region found different approaches to dealing with the arrival of displaced persons, mainly from the Middle East. The migration happened along two main routes. One route goes through Russia to the border with Finland or across the Russian–Norwegian border at Storskog.[2] Here, the Russian authorities only allowed border crossings by vehicle but not on foot (Herrmann, 2015). This measure forced migrants further north to the border between Russia and Norway. A second route led from Sweden to Finland, north from Stockholm, crossing the border at Haparanda–Tornio and then usually south to the Helsinki region (*FTimes*, 2015). More dramatically, in late May 2017, a Ghanaian refugee died at the US–Canadian border from cold exposure (Andrew-Gee, 2017).

The legal situation at the US–Canadian border is similar to the situation in Northern Europe, as the US and Canada have an agreement (Canada and United States, 2004) that requires asylum applications to be processed in the country of first entry (Andrew-Gee, 2017). This makes the situation there similar to the situation within the European Union (Regulation (EU) No. 604/2013). Canada is also a party to the 1951 Refugee Convention, and one result of this combination is that refugees who enter Canada without using regular border crossing points will not be turned back immediately (Andrew-Gee, 2017). This provides an incentive for migrants to try to cross the border from the United States to Canada at places away from regular border crossing points – for example, in forests. This in turn increases the risk of accidents as well as travel times (usually on foot), thereby also increasing the length of time of exposure to cold temperatures and the dangers to human health.

It has been recognized by the UN Refugee Agency (UNHCR) that low temperatures in the wintertime pose a significant risk to displaced persons (Pouilly,

2017). According to the UNHCR's spokesperson, Cécile Pouilly, this problem is by no means restricted to the northernmost parts of Europe (Pouilly, 2017). Indeed, several deaths of displaced persons in early 2017 were attributed to cold temperatures by the UNHCR (Pouilly, 2017). The effect of cold temperatures, combined with insufficient shelter and other services, can be seen in the way migrants were treated in Eastern Europe in the winter of 2016–2017. Dozens of people died owing to the cold, including several refugees (Dearden and McIntyre, 2017). Unlike in Northern Europe, in poorer countries in Southern and Eastern Europe, the available shelter is limited and inadequate for winter conditions, and many refugees are left without shelter and sleep in tents or on the streets (Dearden and McIntyre, 2017). The thousands who lack protection against the cold include "babies, pregnant women and unaccompanied children" (Dearden and McIntyre, 2017: n.p.). Similar situations could arise elsewhere with regard to large numbers of displaced persons. The term "displaced persons" is used in a very wide sense here, including everybody who has lost his or her home, even if only temporarily, be it as a result of armed conflict, natural disaster or other events beyond their control.

Although there is sufficient housing for displaced persons in the Nordic countries, and although the number of persons fleeing to Northern Europe has decreased significantly since 2015, these developments inspired the question of how to protect displaced persons not only in the North, but in general, against the dangers associated with cold temperatures. The question asked here, therefore, is which norms of international human rights law speak to this particular question?

Shelter, understood here as a temporary form of housing,[3] therefore plays an important role in the care provided to displaced persons. The latter term will be used in this text to describe the persons from other countries who are currently seeking to live in Europe, regardless of their specific legal status under international or national law. As a consequence of the increasing number of new arrivals in the region, it has it been necessary for the authorities in Northern Europe to react and to provide sufficient shelter and services for refugees. Although the number of displaced persons coming to the Nordic countries has declined markedly since 2015, which is reflected, for example, in a reduction in asylum applications in Finland (Finnish Immigration Service, 2018), there remains a readiness on the part of the authorities in the region to deal with future increases in refugee numbers. This is particularly the case because the reduction in refugee numbers was due mainly to the deal between the European Union (EU) and Turkey,[4] and the current political situation in Turkey is too unpredictable to make a sufficient assessment. This makes it necessary for all actors in the region that might be involved with refugees or other displaced persons (for example, in the context of a major accident in an Arctic city) in the future to be familiar with the needs of such persons, as well as with their fundamental human rights. In the Arctic, this concerns in particular the right to health in climate and weather conditions for which the persons in question might not be sufficiently equipped.

It is the purpose of this text to identify the international obligations of states when it comes to providing shelter for displaced persons in cold temperature environments and to show that international human rights law can be utilized to safeguard the right of displaced persons to adequate shelter. We begin by describing the health risks associated with exposure to cold (both short-term and long-term exposure). We then analyze the right to health of refugees, looking at different international human rights treaties (with a focus on treaties ratified by Nordic countries) and, finally, we draw conclusions and make recommendations for state action to safeguard the human rights of persons who have been displaced and are exposed to cold temperatures. This is done using literature research in the fields of physiology and international human rights law. In the former case, we rely in particular on PubMed as a search tool. PubMed is made available free of charge by the United States National Library of Medicine. In the context of international human rights law research, emphasis was placed on legal regimes – in particular, international human rights treaties – that apply to the countries of the continental European High North – that is, Norway, Sweden and Finland.

Cold-related health risks

Physiologically, thermal stress is any change in the thermal relation between a temperature regulator and its environment that, if uncompensated by temperature regulation, would result in hyper- or hypothermia (IUPS Thermal Commission, 2001). Thermal comfort is the subjective indifference to the thermal environment. The thermal comfort zone is the range of ambient temperatures, associated with specified mean radiant temperature, humidity and air movement, within which a human in specified clothing expresses indifference to the thermal environment for an indefinite period (IUPS Thermal Commission, 2001).

Weather and environmental conditions have an impact on thermal stress. The cooling effect of air flow (draught), air humidity or wind (wind chill) can increase the effects of cold and has a big influence on human thermal balance (Danielsson, 1996). On the other hand, during physical activity such as walking or cycling, thermal strain can increase, and clothes become wet also in the cold environment. Moist skin and wet clothes are not a good combination in a cold environment. Heat is produced in biochemical processes such as the oxidation of carbohydrates and fat in the human body. Heat is transferred from the inner parts of the body to the body surface by radiation, conduction, convection and vaporization (Hardy and Söderström, 1938). External heat flow from the skin to the environment can lead to the cooling of the body. Thermal strain may become high and increase the thermal stress levels of the body. Accumulated moisture in the clothing not only makes it wet but also causes uncomfortable thermal sensations and poses a health risk in cold conditions (Rintamäki and Rissanen, 2006).

Cold can have harmful effects on health and performance, and other environmental factors can make cold weather even more hazardous. A combination of wind, wet and cold ambient temperature can even lead to hypothermia if the exposure lasts for long and no protective garments are used (Thompson and Hayward, 1996). Cold tolerance differs depending on the person, and individual differences in thermoregulatory responses exist. Hypothermia is widely considered to be a more serious threat for older than younger persons because of older persons' impaired ability to defend body temperature during cold exposure and preventable changes in body composition and physical fitness (Young and Lee, 1997). Cold temperatures can also affect the way pharmaceuticals work in the human body. For example, cold temperatures can lead to changes in the way widely used antihypertensive drugs are absorbed or distributed in or excreted from the body (Komulainen, 2007). This, in turn, means that cold temperatures influence the efficacy of pharmaceuticals. Thermal balance also has effects on human conscious responses and cognitive function (Flouris and Cheung, 2009; Mäkinen et al., 2006; Pääkkönen, 2010; Palinkas et al., 2005). It is important to note that nutritional and hydration statuses have impacts on thermal balance, both in hot and cold environments, and, therefore, dehydration of the body may expose people to injuries with both cases decreasing physical and cognitive performance (Marriott and Carlson, 1996).

Low ambient temperatures can expose individuals to severe health risks (Biem et al., 2003). Fatal hypothermia can happen even in relatively mild temperatures. Most mortality cases have been found to happen when the temperature of the environment is above 0° Celsius. According to physiological studies, an air temperature of 25°C exposes us to hypothermia. In water, the risk limit for hypothermia is even greater, being as high as 33°C (Brändström et al., 2012). Cold tolerance differs depending on the person and individual variation in thermoregulatory responses (Brändström et al., 2012). Our body temperature is regulated to remain close to a set point of 37°C. According to Parsons, the core temperature of the human body can vary safely only by 4.5°C (Parsons, 2003). Physiological functions in humans work properly in the zone between hypothermia (35.0°C) and hyperthermia (39.5°C). The thermo-neutral ambient temperature for naked, resting humans is approximately 27°C.

This means that this is the optimal temperature for humans to avoid body cooling: below 27°C, one has to move, or use another physiological mechanism such as burning brown body fat to produce body heat, or use external heat sources. Above 27°C, heat loss by evaporation begins, resulting in cooling (IUPS Thermal Commission, 2001). Thermal adaptation to cold (acclimatization or acclimation) takes approximately 2 weeks (IUPS Thermal Commission, 2001). Newly arrived displaced persons who are not used to cold temperatures often are unable to acclimatize fast enough to avoid health risks. Humidity and skin moisture increase the rate of body cooling (Holmer, 1993). This risk is particularly relevant if adequate shelter is missing.

Physiological stress induced by excessive cold can impair functioning of the body (Biem et al., 2003). Skin, hand and foot cooling can cause discomfort, pain or even cold injuries (Rintamaki et al., 1993). As was shown by Pääkkönen, cold can change our mood and cognition, and it has effects on the pineal and thyroid hormones (Pääkkönen, 2010). Cold can lead to accidental hypothermia or cold injuries if the risks are not observed and taken into account. Cold can have effects on general health or it can worsen the symptoms of existing medical disorders, such as diabetes mellitus (see e.g. Parsons, 2003; Rintamäki, 2007; Scott, Bennett, and Macdonald, 1987).

Cold can pose a risk to persons with specific health conditions. Problems with the circulation in hands and feet, previous frostbite, coronary artery disease, severe hypertension, severe symptoms in joints and muscles, severe infections, trauma, respiratory failure or some endocrine disorders may all cause increased sensitivity to cold (Biem et al., 2003). Also malnutrition increases the risk of hypothermia (Young et al., 1998). In general, it has to be noted that children and the elderly are particularly vulnerable because they are less capable of maintaining body temperature than adults under the age of 60 (Castellani and Young, 2016; Castellani et al., 2006; Young and Lee, 1997). In children, this is owing to the ratio of body surface to body mass, which means that they are likely to lose body heat quickly (Young and Lee, 1997). Children and older persons are likely to have difficulties when it comes to taking measures to protect themselves against cold (Castellani et al., 2006; Frank et al., 2000). The mortality risk is increased – for example, in cases of long-term cold exposure (e.g. in the case of homelessness) or substance abuse.

The rights of displaced persons to health and housing

Universal Declaration of Human Rights

Although civil and political rights have long dominated public human rights discourse, international human rights law has concerned itself with social rights since the beginning of the discipline in the modern sense of the term. Although technically non-binding, and hence depending on the good will of states when it comes to enforcement, Article 25 (1) of the Universal Declaration of Human Rights (UDHR) guarantees for everybody – that is, citizens and non-citizens – among other rights, "the right to a standard of living adequate for the health and well-being of himself and of his family, including food, clothing, housing and medical care and necessary social services" (Article 25 (1) UDHR). The UDHR has been adopted in the form of a resolution of the General Assembly of the United Nations. This makes the UDHR technically non-binding, but the UDHR has, through decades of state practice and *opinio juris* (which are the essential elements for the customary formation of international law), long become, in the words of Hannum, "significant evidence of customary international law" (Hannum, 1995/96). Everybody has the right to complain about violations of the UDHR to the Human Rights Council of the United Nations (United Nations, 2017a). In practice, though, the enforcement of rights under the UDHR requires the cooperation of the states concerned.

General human rights treaties

In addition to the UDHR, human rights standards are included in a wide range of international treaties, many of which have their own complaint mechanisms.[5] This led to the development of the International Covenant on Civil and Political Rights (ICCPR) and the International Covenant on Economic, Social and Cultural Rights (ICESCR), at the global level, and to regional human rights conventions.

The first of the latter was the European Convention on Human Rights (ECHR), to which all states in the European Arctic are parties (Council of Europe, 2017a). Originally concerned with civil and political rights, the ECHR has been supplemented over recent decades with protocols that wide its scope. In addition, the European Social Charter (ESC) contains social rights. Both the ECHR and the ESC were drafted by the Council of Europe, not the European Union (of which Russia and Norway are not members), and all four states in the region are parties to the ECHR (Council of Europe, 2017b), the ESC (Council of Europe, 2017c), the ICCPR (United Nations, 2017b) and the ICESCR (ibid.).

Article 6 (1) ICCPR and Article 2 (1) ECHR protect the right to life. Although the wording of the norm indicates a classical view of human rights norms as preventing the state from taking actions that would harm human rights, the so-called "negative dimension of human rights" (Plesner, 2001: 4), especially the right to life, is now commonly understood to have a positive dimension as well (Akandji-Kombe, 2007: 21). States, therefore, are required to take the actions necessary to preserve human life – for example, through enacting laws and enforcing them, but also through the maintenance of necessary infrastructure, such as a functioning health care system. Under these norms, states are required to protect human life. Article 11 (1) ICESCR is more specific in that it obliges states to take measures towards the full realization of everybody's right "to an adequate standard of living for himself and his family, including adequate food, clothing and housing, and to the continuous improvement of living conditions" (Article 11 (1) sentence 1 ICESCR). Like other social rights in the ICESCR, the rights contained in Article 11 ICESCR are meant to be implemented over time, as is emphasized by Article 2 (1) ICESCR.

This progressive realization of social human rights is both a challenge and a chance for poorer countries. In states that are as developed as the Nordic countries, adequate shelter that at least protects against the predictable effects of winter weather and the associated temperatures can reasonably be expected to be (and usually is)[6] provided by the authorities; the question is under which international human rights norms this moral obligation actually becomes an international legal obligation. Once the existence of such a legal obligation has been confirmed, it follows – for example, from Articles 2 (1) and 1 (1) of the Convention on the Elimination of all Forms of Racial Discrimination (CERD) – that services and infrastructure, including shelter, must be provided without discrimination "based on race, colour, descent, or national or ethnic origin" (Article 1 (1) CERD). In addition to Article 14 ECHR, which requires non-discrimination in the application of the ECHR, Article 1 (1) of Protocol 12 to the ECHR includes a wider prohibition of discrimination, which applies to Finland but not to the other states under consideration here (Council of Europe, 2017d):

> The enjoyment of any right set forth by law shall be secured without discrimination on any ground such as sex, race, colour, language, religion, political or other opinion, national or social origin, association with a national minority, property, birth or other status.
>
> (Article 1 Protocol 12 to the ECHR)

The non-discrimination approach to implementing social rights is especially important in light of tendencies in the domestic politics of some European states as anti-immigrant sentiment has grown in the wake of the 2015 immigrant crisis. This has been reflected not only in political gains for far-right parties, but also in discussions concerning potential limits of social services for migrants (Tharoor, 2015). It follows from the duty not to discriminate that health policies have to include everybody (cf. Aboii, 2014; Hodge et al., 2016). Raising awareness of the legal duty not to discriminate when it comes to securing human rights and of the international law dimension of the non-discrimination obligation is an important aspect of implementing the right to shelter effectively.

Specialized human rights treaties

In the following, it will be shown that international human rights provide a multitude of norms that provide a legal basis for an obligation to provide shelter. In addition to the aforementioned general international human rights treaties, there are also more specialized international human rights treaties that, like the CERD, deal with particular issues and special groups of persons. The most obvious example in this regard is the Refugee Convention (RC). This international treaty dates back to 1951. Although originally intended to deal with the situation in the wake of World War II, the RC has become, and remains, the most important set of international norms dealing with refugees. According to Article 21 RC, states that are parties to the RC (including the four states under consideration here) shall treat refugees no differently from other foreigners lawfully residing in the country when it comes to providing shelter. The norm is particular in that it does not provide for a right to shelter, but prohibits discrimination in the application of domestic laws related to shelter (United Nations, 2017c). This follows from the phrase "in so far as the matter is regulated by laws or regulations or is subject to the control of public authorities" in Article 21 RC. Article 22 (1) of the Convention on the Rights of the Child (CRC) provides for a duty of state to provide humanitarian assistance to children.

Article 3 (1) of the CRC requires "public or private social welfare institutions, courts of law, administrative authorities [and] legislative bodies" to give most attention to the interests of the child whenever an action is taken that concerns a child. These interests include the human rights of children, such as the right to life, the right not to be separated from their parents, the right to contact with parents who live in a different country, and the right to health (Article 3 (1) CRC; Article 6 (1) CRC; Article 9 (1) sentence 1 CRC; Article 10 (2) sentence 1 CRC; Article 24 CRC). The latter right is protected by Article 24 CRC in the form of a right to services and infrastructure – specifically, "the right of the child to the enjoyment of the highest attainable standard of health and to facilities for the treatment of illness and rehabilitation of health" (Article 24 (1) sentence 1 CRC). Accordingly, states "shall strive to ensure that no child is deprived of his or her right of access to such health care services" (Article 24 (1) sentence 2 CRC). Similarly, Article 27 CRC provides for the "right of every child to a standard of living adequate for the child's physical, mental, spiritual, moral and social development" (Article 27 (1) CRC). Although the obligation to realize this right rests first and foremost with the parents, states "shall take appropriate measures to assist parents and others responsible for the child to implement this right and shall in case of need provide material assistance and support programs, particularly with regard to nutrition, clothing and housing" (Article 27 (2) CRC; Article 27 (3) CRC). This, too, contributes to the obligation of states to provide shelter if needed.

Article 42 CRC obliges the states parties to the Convention to disseminate knowledge of the rights of children as enshrined in the CRC as widely as possible. It follows from Article 42 CRC, therefore, that there is a separate obligation of states parties to the CRC (including the four countries under discussion here) to make the obligations under the CRC widely known. This, in turn, enables individuals, including displaced persons as well as decision-makers at the local level, to be aware of the rights of children under the CRC. Raising awareness is an important step towards implementing the right to shelter.

Other specialized human rights treaties can be taken into account in this context as well. For example Article 14 of the Convention on the Elimination of All Forms of Discrimination Against Women (CEDAW) provides for a duty of states to ensure "adequate living conditions, particularly in relation to housing, sanitation, electricity and water supply, transport and communications" (Article 14 (2) (h) CEDAW). Although Article 14 CEDAW refers specifically to women in rural areas and appears to relate mainly to women living in their home communities, it also applies to female displaced persons in the thinly populated region discussed here.

Another example of a contribution to a trend towards the emergence of universal (customary law) right to shelter is the Convention on the Rights of Persons with Disabilities (CRPD), which has been ratified by all four states in the region, and Finland and Sweden have also ratified the Optional Protocol thereto (CRPD-OP; see United Nations, 2016). As in the case of children, pregnant or breastfeeding women, the elderly, persons with diabetes or other pre-existing (or even unknown) health issues, cold exposure can become a health threat for persons with disabilities. It is, therefore, vital that the rights of particularly vulnerable groups are taken into account by decision-makers at the local level. At the international level, as has been shown, there already exist a range of norms that provide some degree of protection. These norms include specific rules in both general and special international human rights law that oblige states to provide adequate shelter. In order for these norms to be truly effective, though, it is necessary that states ensure their implementation not just on paper, but in practice, which requires cooperation with all levels of public authority, dissemination of knowledge of these obligations and often significant investments in infrastructure, material and staff.

In Article 10 CRPD, the

> States Parties reaffirm that every human being has the inherent right to life and [the states that have ratified the CRPD accept the obligation that they]

shall take all necessary measures to ensure its effective enjoyment by persons with disabilities on an equal basis with others.

(United Nations, 2016)

In addition, Article 25 sentence 1 CRPD protects "the right to the enjoyment of the highest attainable standard of health without discrimination on the basis of disability" (Article 25 sentence 1 CPRD, United Nations, 2016). The phrase "the highest attainable standard" (ibid.) reflects an approach that is common in social human rights in that every state has to do whatever it is capable of, given its specific situation. This allows states to fulfil their obligations step by step. On the other hand, states are required to take some action and to make progress. In the case of persons with disabilities, Article 25 sentence 1 CRPD reiterates an obligation that already exists under Article 12 (1) ICESCR, which protects "the right of everyone to the enjoyment of the highest attainable standard of physical and mental health" (Article 12 (1) CRPD).

In addition, Article 26 (1) CRPD obliges the state to ensure adequate shelter. Although the norm does not necessarily provide a subjective right to shelter, Article 26 CRPD is noteworthy in that it outlines, in significant detail, which kinds of measures have to be taken by states in order to improve the situation of persons with disabilities. Similarly, Article 28 CRPD describes possible measures that "shall" be taken by states in order to fully realize the right of persons with disabilities "to an adequate standard of living for themselves and their families, including adequate food, clothing and housing, and to the continuous improvement of living conditions" (Article 28 (2) CRPD; Article 28 (1) CRPD).

In the context of the situation described at the beginning of this text, particular attention should be given to Article 11 CRPD, which deals with "situations of risk, including situations of armed conflict, humanitarian emergencies and the occurrence of natural disasters" (Article 11 CRPD). Referring to states' pre-existing "obligations under international law, including international humanitarian law and international human rights law", the norm is a reminder that persons with disabilities also have a right to safety and protection by the state. These rights exist independent of Article 11 CRPD, which is only a reminder, albeit a legally binding reminder of existing obligations of states in times of crisis. What is important is that, like many other social rights in international law, this right to safety and protection applies in crisis situations – that is, in the very kind of situation in which governments might otherwise be tempted to derogate from human rights obligations, in as far as international human rights treaties permit states to do so (Article 11 CRPD; Kirchner, 2010).

Although some of the persons migrating to the continental European Arctic states might be migrant workers, the Migrant Workers Convention (MWC) is not applicable there, as it has not been ratified by any of the states in question. On states that have indeed ratified the MWC, Articles 28 and 45 (1) (c) MWC impose an obligation to provide access to social and health care services for workers and their family members, regardless of their residence status or work permission under domestic law. This can include access to emergency housing.

Conclusions

International human rights law is implemented first and foremost at the national level. Although international human rights litigation can be one way to force authorities to take action, it is a slow way to achieve change. From a practical perspective, raising knowledge and awareness of human rights obligations among decision-makers on the local level is an essential step towards increasing the effectiveness of human rights implementation. When it comes to providing shelter for displaced persons in cold climates, awareness among local decision-makers of the particular health risks faced by displaced persons who are not familiar with cold climates is the key to effective human rights implementation at the local level. This has to include awareness of the legal nature of the obligations discussed here.

Exposure to cold temperatures poses serious health threats, and effective protection of displaced persons in cold climates creates particular challenges for authorities and volunteers. International human rights law provides for a right to health and, if necessary, also requires the state to provide shelter. The right to shelter is a functional part of the human right to health, which is, directly or indirectly, protected through a number of human rights treaties. The realization of these human rights has to be effective and meaningful. In light of the aforementioned winter temperatures experienced in south-eastern Europe, it has to be concluded that the measures taken there have not been sufficient to protect refugees, and that they are not compatible with international law.

Overall, minimum standards of care have to be respected. This includes the provision of infrastructure that is sufficient to prevent harm to human health. As far as these standards follow from the ICESCR, they do not merely describe a minimum, but they require the "highest attainable standard of health" (Committee on Economic, Social and Cultural Rights, 2000). The use of the term "highest attainable standard" (Article 12 (1) ICESCR) in the ICESCR highlights that these obligations do not need to be fulfilled overnight, but that they can be implemented step by step, improving over time.

This obligation takes into account both individual characteristics as well as the ability of the state to comply with the obligations under Article 12 ICESCR (Committee on Economic, Social and Cultural Rights, 2000, para. 9). In the relatively wealthy countries that are researched here, this means a higher level of obligation. This is particularly the case as shelter, both for residents and newcomers, is widely available in general, at least in the wealthier European Arctic nations of Finland, Sweden and Norway. The ability to provide shelter that is sufficient to safeguard human health translates into an obligation to provide sufficient shelter to displaced persons as well – to the best ability of the states. As far as these standards are required by the other human rights treaties described here, they form obligations that have to be implemented immediately. This means that states have to take precautions for the event of a sudden need for shelter for large numbers of persons during wintertime, for example, as is the case in Finland, by maintaining container shelters in case the existing predesignated buildings should prove to be insufficient. This obligation applies to all persons concerned, including locals or tourists affected by a natural disaster or other mass casualty event, as well as refugees or other displaced persons.

Notes

1 In this text, the term "displaced persons" is used to refer to migrants, refugees etc. who have found themselves in the geographical region that is the focus of this research. Because this text is concerned with matters that are not dependent on the legal status of the persons concerned, the terms "displaced persons", "refugees" and "migrants" are used interchangeably.
2 Russia and Belarus form a supranational union, the Union State, which seems to have limited practical effect. The 1998 Treaty on Equal Rights of Citizens between Belarus and Russia guarantees cross-border access to employment and health care. On the situation in light of the ongoing Ukraine war, see Anishchanka (2015). See also Herrmann (2015) and Staalesen (2016).
3 On housing as a human right, see Adams (2009) and Hartman (1998).
4 Technically, the agreement between the European Union and Turkey was a joint statement that was published as a press release (EU and Turkey, 2016). See the addition to the press release (ibid.) for cases in which the EU's General Court (formerly known as the Court of First Instance) made reference to this press release.
5 For an overview over litigation options see, e.g., Kirchner (2018).
6 Particularly noteworthy is Finland's success in the prevention of homelessness, based on a "Housing First" approach (Finnish Ministry of the Environment, 2016: 2).

References

Aboii, S.M. (2014). "Undocumented immigrants and the inclusive health policies of sanctuary cities", *Harvard Public Health Review*, http://harvardpublichealthreview.org/undocumented-immigrants-and-the-inclusive-health-policies-of-sanctuary-cities/

Adams, K.D. (2009). "Do we need a right to housing?", *Nevada Law Journal*, Vol. 9, pp. 275–324.

Akandji-Kombe, J.-F. (2007). *Positive obligations under the European Convention on Human Rights*, www.echr.coe.int/LibraryDocs/DG2/HRHAND/DG2-EN-HRHAND-07(2007).pdf

Andrew-Gee, E. (2017). "With asylum seeker's reported death, Canada–U.S. crossing debate intensifies", *The Globe and Mail*, 31 May 2017, www.theglobeandmail.com/news/national/with-asylum-seekers-reported-death-canada-us-crossing-debate-intensifies/article35171898/

Anishchanka, M. (2015). "Is Belarus and Russia's 'brotherly love' coming to an end?" *The Guardian*, 29 May 2015, www.theguardian.com/world/2015/may/28/belarus-russia-brotherly-love-ukraine-crisis

Biem, J., Koehncke, N., Classen, D., Koehncke, N., Classen, D., & Dosman, J. (2003). "Out of the cold: management of hypothermia and frostbite", *Canadian Medical Association Journal*, Vol. 168, No. 3, pp. 305–311.

Brändström, H., Eriksson, A., Giesbrecht, G., Angquist, K.A., & Haney, M. (2012). "Fatal hypothermia: an analysis from a sub-arctic region", *International Journal of Circumpolar Health*, Vol. 71, pp. 1–7. doi:10.3402/ijch.v71i0.18502

Canada and United States. (2004). *Agreement between the Government of Canada and the Government of the United States of America for cooperation in the examination of refugee status claims from nationals of third countries, Canada Treaty Series 2004/2*, www.treaty-accord.gc.ca/details.aspx?id=104943

Castellani, J.W., & Young, A.J. (2016). "Human physiological responses to cold exposure: acute responses and acclimatization to prolonged exposure", *Autonomic Neuroscience: Basic and Clinical*, Vol. 196, pp. 63–74.

Castellani, J.W., Young, A.J., Ducharme, M.B., Giesbrecht, G.G., Glickman, E., & Sallis, R.E. (2006). "Prevention of cold injuries during exercise", *Medicine & Science in Sports & Exercise*, Vol. 38, No. 11, pp. 2012–2029. doi:10.1249/01.mss.0000241641.75101.64

Committee on Economic, Social and Cultural Rights. (2000) *General Comment No. 14, The right to the highest attainable standard of health (article 12 of the International Covenant on Economic, Social and Cultural Rights)*, E/C.12/2000/4, 11 August 2000, http://docstore.ohchr.org/SelfServices/FilesHandler.ashx?enc=4sl Q6QSmlBEDzFEovLCuW1AVC1NkPsgUedPlF1vfPMJ2c7ey6PAz2qaoj TzDJmC0y%2B9t%2BsAtGDNzdEqA6SuP2r0w%2F6sVBGTpvTSCbi Or4XVFTqhQY65auTFbQRPWNDxL

Convention for the Protection of Human Rights and Fundamental Freedoms, European Treaty Series No. 5, European Convention on Human Rights, as amended by Protocols 11 and 14, 4 November 1950, www.echr.coe.int/Documents/Convention_ENG.pdf.

Convention on the Elimination of All Forms of Discrimination against Women, United Nations General Assembly Resolution 34/180, 18 December 1979, www.ohchr.org/ Documents/ProfessionalInterest/cedaw.pdf

Convention on the Rights of Persons with Disabilities, www.ohchr.org/EN/HRBodies/ CRPD/Pages/ConventionRightsPersonsWithDisabilities.aspx

Convention on the Rights of the Child, United Nations General Assembly Resolution 44/25, 20 November 1989, www.ohchr.org/Documents/ProfessionalInterest/crc.pdf

Convention Relating to the Status of Refugees, 1951, reprinted in: UNHCR, Convention and Protocol Relating to the Status of Refugees, pp. 6 et seq, www.unhcr.org/3b66c2aa10.pdf

Council of Europe. (2017a). *Simplified Chart of Signatures and Ratifications*, Status as of 05/02/2017, Selected subject-matters: Human Rights (Convention and Protocols only), www.coe.int/en/web/conventions/search-on-treaties/-/conventions/chartSignature/3

Council of Europe. (2017b). *Chart of signatures and ratifications of Treaty 005, Convention for the Protection of Human Rights and Fundamental Freedoms*, Status as of 05/02/2017, www.coe.int/en/web/conventions/full-list/-/conventions/treaty/005/signatures

Council of Europe. (2017c). *European Social Charter, Signatures & Ratifications*, www.coe. int/en/web/turin-european-social-charter/signatures-ratifications

Council of Europe. (2017d). *Chart of signatures and ratifications of Treaty 177, Protocol No. 12 to the Convention for the Protection of Human Rights and Fundamental Freedoms*, Status as of 05/02/2017, www.coe.int/en/web/conventions/full-list/-/conventions/ treaty/177/signatures?p_auth=SlVGIAdM

Danielsson, U. (1996). "Windchill and a risk of tissue freezing", *Journal of Applied Physiology*, Vol. 81, No. 6, pp. 2666–2673. doi:10.1152/jappl.1996.81.6.2666

Dearden, L., & McIntyre, N. (2017). "Refugees freezing to death across Europe after 'continued failure' on crisis leaves thousands at risk – Two Iraqi men were found dead after walking for 48 hours through heavy snow", *The Independent*, 11 January 2017, www.independent.co.uk/news/world/europe/europe-refugees-freeze-to-death-hypothermia-bulgaria-athens-cold-weather-serbia-sleeping-rough-a7520106.html

EU and Turkey. (2016). *European Council/Council of the European Union, EU–Turkey statement*, 18 March 2016, www.consilium.europa.eu/en/press/press-releases/ 2016/03/18/eu-turkey-statement/

European Social Charter, European Treaty Series No. 163, 3 May 1996, https:// rm.coe.int/CoERMPublicCommonSearchServices/DisplayDCTMContent? documentId=090000168007cf93

Finnish Immigration Service. (2018). "Statistics for 2017: clearly less asylum seekers than the year before – over 2,100 asylum seekers submitted their first application", 30 January 2018, https://migri.fi/en/article/-/asset_publisher/vuoden-2017-tilastot-turvapaikanhakijoita-selvasti-edellisvuosia-vahemman-ensimmaisen-hakemuksen-jatti-reilut-2-100-hakijaa

Finnish Ministry of the Environment. (2016). *Action Plan for Preventing Homelessness in Finland 2016–2019*, Decision of the Finnish Government, 9 June 2016, http://asuntoensin.fi/assets/files/2016/11/ACTIONPLAN_FOR_PREVENTING_HOMELESSNESS_IN_FINLAND_2016_-_2019_EN.pdf

Flouris, A.D., & Cheung, S.S. (2009). "Human conscious response to thermal input is adjusted to changes in mean body temperature", *British Journal of Sports Medicine*, Vol. 43, No. 3, pp. 199–203.

Frank, S.M., Raja, S.N., Bulcao, C., & Goldstein, D.S. (2000). "Age-related thermoregulatory differences during core cooling in humans", *American Journal of Physiology – Regulatory, Integrative and Comparative Physiology*, Vol. 279, pp. R349–R354.

FTimes. (2015). "Finland to control refugee entry from Sweden", *Finland Times*, 15 September 2015, www.finlandtimes.fi/national/2015/09/15/20436/Finland-to-control-refugee-entry-from-Sweden.

Hannum, H. (1995/96). "The status of the universal declaration of human rights in national and international law", *Georgia Journal of International and Comparative Law (1995/96)*, 25, pp. 287–397, at p. 322.

Hardy, J.D., & Söderström, G.F. (1938). "Heat loss from the nude body and peripheral blood flow at temperatures of 22°C to 35°C", *Journal of Nutrition*, Vol. 16, No. 5, pp. 493–510.

Hartman, C. (1998). "The case for a right to housing", *Housing Policy Debate*, 9, pp. 223–246.

Herrmann, G. (2015). "Flüchtlinge in der Arktis - Radeln in den Schengenraum", *Süddeutsche Zeitung*, 24 October 2015, www.sueddeutsche.de/politik/fluechtlinge-in-der-arktis-radeln-in-den-schengenraum-1.2706840

Hodge, J.G., Jr., Weidenaar, K., Baker-White, A., Barraza, L., Crock Bauerly, B., Corbett, A., Davis, C., Frey, L.T., Griest, M.M., Healy, C., Krueger, J., McGowan Lowrey, K., & Tilburg, W. (2016). Legal innovations to advance a culture of health. *Journal of Law, Medicine and Ethics*, 43, 904–912. doi:10.1111/jlme.12328

Holmer, I. (1993). "Work in the cold. Review of methods for assessment of cold exposure", *Archives of Occupational & Environmental Health*, Vol. 65, pp. 147–155.

Institute of Medicine (US) Committee on Military Nutrition Research, Marriott, B.M., & Carlson, S.J. (Eds.). (1996). *Nutritional Needs in Cold and in High-Altitude Environments: Applications for Military Personnel in Field Operations*. National Academies Press (US); 9, Influence of Cold Stress on Human Fluid Balance, Washington, DC. www.ncbi.nlm.nih.gov/books/NBK232870/

International Convention on the Elimination of All Forms of Racial Discrimination, United Nations General Assembly Resolution 2106 (XX), 21 December 1965, www.ohchr.org/Documents/ProfessionalInterest/cerd.pdf

International Convention on the Protection of the Rights of All Migrant Workers and Members of Their Families, United Nations General Assembly Resolution 45/158, 18 December 1990, www.ohchr.org/EN/ProfessionalInterest/Pages/CMW.aspx

International Covenant on Civil and Political Rights, United Nations General Assembly Resolution 2200A (XXI), 16 December 1966, www.ohchr.org/Documents/ProfessionalInterest/ccpr.pdf

International Covenant on Economic, Social and Cultural Rights, United Nations General Assembly Resolution 2200A (XXI), 16 December 1966, www.ohchr.org/Documents/ProfessionalInterest/cescr.pdf

IUPS Thermal Commission. (2001). "Glossary of terms for thermal physiology", 3rd ed., Japanese, *Journal of Physiology*, Vol. 51, pp. 245–280.

Kirchner, S. (2010). "Human rights guarantees during states of emergency – the European Convention on Human Rights", *Baltic Journal of Law and Politics*, Vol. 3, No. 2, pp. 1–25.

Kirchner, S. (2018). "Lis alibi pendens in International Human Rights Litigation", *Edilex Lakikirjasto*, 2 March 2018, www.edilex.fi/artikkelit/18545.pdf

Komulainen, S. (2007). *Effect of antihypertensive drugs on blood pressure during exposure to cold. Experimental study in normotensive and hypertensive subjects*. PhD thesis. Acta Universitatis Ouluenisis. University of Oulu, Oulu, Finland.

Mäkinen, T.M., Palinkas, L.A., Reeves, D.L., Pääkkönen, T., Rintamäki, H., Leppäluoto, J., & Hassi, J. (2006). "Effect of repeated exposures to cold on cognitive performance in humans", *Physiological Behavior*, Vol. 87, No. 1, pp. 166–176. doi:10.1016/j. physbeh.2005.09.015

Optional Protocol to the Convention on the Rights of Persons with Disabilities, www.ohchr. org/EN/HRBodies/CRPD/Pages/OptionalProtocolRightsPersonsWithDisabilities. aspx

Pääkkönen, T. (2010). *Melatonin and thyroid hormones in the cold and in darkness. Association with mood and cognition*. PhD thesis Acta Universitatis Ouluensis. University of Oulu, Oulu, Finland.

Palinkas, L.A., Mäkinen, T.M., Pääkkönen, T., Rintamäki, H., Leppäluoto, J., & Hassi, J. (2005). "Influence of seasonally adjusted exposure to cold and darkness on cognitive performance in circumpolar residents", *Scandinavian Journal of Psychology*, Vol. 46, pp. 239–246.

Parsons, K. (2003). *Human Thermal Environments: The Effects of Hot, Moderate and Cold Environments on Human Health, Comfort and Performance* (2nd ed.). Taylor & Francis CRC Press, London.

Plesner, I.T. (2001). "State Church and Church Autonomy in Norway", in G. Robbers (Ed.), *Church Autonomy: A Comparative Survey* (1st ed.). Frankfurt am Main: Peter Lang Verlag. www.strasbourgconsortium.org/content/blurb/files/Chapter%2022.%20 Plesner.pdf

Pouilly, C. (2017). "Refugees and migrants face high risks in winter weather in Europe", *UNHCR Briefing Notes*, 13 January 2017, www.unhcr.org/news/ briefing/2017/1/58789f624/refugees-migrants-face-high-risks-winter-weather-europe.html

Protocol No. 12 to the Convention for the Protection of Human Rights and Fundamental Freedoms, European Treaty Series No. 177, www.coe.int/en/web/conventions/ full-list/-/conventions/treaty/177

Regulation (EU) No. 604/2013 of the European Parliament and of the Council of 26 June 2013 establishing the criteria and mechanisms for determining the Member State responsible for examining an application for international protection lodged in one of the Member States by a third-country national or a stateless person, Official Journal L 180, 29 June 2013, pp. 31–59, http://data.europa.eu/eli/reg/2013/604/oj

Rintamäki, H. (2007). "Human responses to cold", *Alaska Medicine*, Vol. 49, No. 2, pp. 29–31.

Rintamaki, H., Hassi, J., Smolander, J., Louhevaara, V., Rissanen, S., Oksa, J., & Laapio, H. (1993). "Responses to whole body and finger cooling before and after an Antarctic expedition", *European Journal of Applied Physiology*, Vol. 67, No. 4, pp. 380–384.

Rintamäki, H., & Rissanen, S. (2006). "Heat strain in cold", *Industrial Health*, Vol. 44, pp. 427–432.

Scott, A.R., Bennett, T., & Macdonald, I.A. (1987). "Diabetes mellitus and thermoregulation", *Canadian Journal of Physiology and Pharmacology*, Vol. 65, pp. 1365–1376.

Staalesen, A. (2016). "End comes to special regime on Finnish–Russian border", *Barents Observer*, 8 September 2016, https://thebarentsobserver.com/en/borders/2016/09/end-comes-special-regime-finnish-russian-border

Tharoor, I. (2015). "Europe's refugee crisis strengthens far-right parties", *Washington Post*, 13 October 2015, www.washingtonpost.com/news/worldviews/wp/2015/10/13/europes-refugee-crisis-strengthens-far-right-parties/

Thompson, R.L., & Hayward, J.S. (1996). "Wet-cold exposure and hypothermia: thermal and metabolic responses to prolonged exercise in rain", *Journal of Applied Physiology*, Vol. 81, pp. 1128–1137.

United Nations. (2016). *CRPD and Protocol Signatures and Ratification*, Map No. 4496 Rev. 6, May 2016, www.un.org/disabilities/documents/2016/Map/DESA-Enable_4496R6_May16.jpg

United Nations. (2017a). "Human rights, office of the high commissioner", *Human Rights Council Complaint Procedure*, www.ohchr.org/EN/HRBodies/HRC/ComplaintProcedure/Pages/HRCComplaintProcedureIndex.aspx

United Nations. (2017b). "Human rights, office of the high commissioner", *Status of Ratification Interactive Dashboard*, http://indicators.ohchr.org/

United Nations. (2017c). High commissioner for refugees, *States Parties to the 1951 Convention relating to the Status of Refugees and the 1967 Protocol*, www.unhcr.org/protect/PROTECTION/3b73b0d63.pdf

Universal Declaration of Human Rights, United Nations General Assembly Resolution A/RES/3/217 A, 10 December 1948, www.un.org/en/universal-declaration-human-rights/index.html

Young, A.J., Castellani, J.W., O'Brien, C., Shippee, R.L., Tikuisis, P., Meyer, L.G., Blanchard, L.A., Kain, J.E., Cadarette, B.S., & Sawka, M.N. (1998). "Exertional fatigue, sleep loss, and negative energy balance increase susceptibility to hypothermia", *Journal of Applied Physiology*, Vol. 85, pp. 1210–1217.

Young, A.J., & Lee, D.T. (1997). "Aging and human cold tolerance", *Experimental Aging Research*, Vol. 23, pp. 45–67.

Part V

Migration and development issues in the Arctic

10 Mixed embeddedness of immigrant entrepreneurs and community resilience

Lessons for the Arctic

Jan Brzozowski

Introduction

There is no doubt that the Arctic is facing serious population and economic challenges. Apart from already being the most sparsely populated world region, the Arctic has experienced an additional decrease in the overall population in recent decades (Hamilton & Mitiguy, 2009). Between 2000 and 2013, the total population of the Arctic dropped by 56,000 (and by 1.4 per cent; cf. Heleniak & Bogoyavlensky, 2015). Yet, there are some prospects of reversing this negative demographic trend and enhancing population growth in the Circumpolar North. The most obvious change is the gradual warming in the Circumpolar North. Just to take the example of the Svalbard archipelago, mean annual temperatures have risen by 2°C since the 1980s, resulting in a decrease in the area of local glaciers and their overall retreat (Ziaja et al., 2016). Owing to climate change and global warming, the natural resources of the region are becoming much more accessible, and their economic exploitation is becoming both technologically possible and profitable.

But these changes are to be even more intense. For instance, a recent study on climate change for northern Asian Russia predicts much a milder climate for this region by the 2080s and a significant shift of the permafrost to the north-east. This means that substantial parts of the Russian Arctic would move from climate severity categories "extreme" and "absolutely extreme" to "unfavourable" or even "moderately favourable" (Parfenova et al., 2019). As a result, in spite of the current depopulation problems of the region (Heleniak, 2012),[1] the Russian Arctic could become an attractive destination for migration in the upcoming decades.

Consequently, not only Russia (Romashkina et al., 2017), but all of the countries of the Arctic perceive the region as strategic, aiming to increase its population in order to exploit its future economic potential. Yet, the main barrier to the future economic development is precisely the sparsity of the Arctic population: currently, in most of the Arctic subregions, the local population is so small that the costs of the public services are extremely high (Brooks & Frost, 2012). Moreover, the small population size results in small local markets and poor infrastructure, which in turn has a negative effect on the possibilities of economic advancement of locals and newcomers. At the same time, the rising sea level and permafrost thaw

(Ziaja et al., 2016) can harm indigenous populations and the environment, leading to forced population movements in the nearest future. This means that most of the Arctic subregions are somehow trapped in a vicious circle of peripherality: small population size combined with relatively poor infrastructure; domination of the large-scale extraction of non-renewable resources, which negatively affects environment, but creates little positive spillovers for local societies (Duhaime & Caron, 2006); and low levels of R&D and innovation (Nilsen, 2016) resulting in relative underdevelopment, which in turn reduces the attractiveness of the Arctic for new immigrants – both internal and international ones.

Therefore, the Arctic region countries need to develop an environment that is friendly and attractive to newcomers and that would facilitate the socio-economic development of the region. This environment has to be characterized by community resilience – that is, the ability to positively respond to and even influence social and economic change, while keeping the essence of the local cultural, social and economic integrity within critical thresholds (Berkes & Ross, 2013). This means being open to new ethnic groups: their costumes, needs and cultural norms, but at the same time preserving and praising the existing culture, especially in the case of indigenous populations (see the chapter by Merhar, 2020, in this volume). As the "new" immigration to the Arctic is a very recent phenomenon, it is very important for the regional policymakers to capitalize on the experience of other, more established countries of immigration in order to learn from their success stories, but also from their mistakes and problems. In this case, the mixed embeddedness concept developed by Kloosterman and Rath (2001) is an interesting theoretical approach that could serve as an inspiration for building community resilience towards immigrants and their entrepreneurial activity. This theory provides a set of propositions for both researchers in migration studies and policymakers interested in supporting immigrant entrepreneurship.[2] As, in the Arctic, entrepreneurial rates and innovative activities are relatively low (Exner-Pirot, 2015; Krasulina, 2018), this concept can be very useful for enhancing the community resilience in the region.

The aim of this chapter is to present the concept of mixed embeddedness of immigrant entrepreneurship and integrate it with the theory of community resilience within the Arctic/Circumpolar North context. Therefore, the chapter conducts both conceptual and case-study analyses in order to offer some insights into the main policy questions: how to attract immigrants into the Arctic, and how to make the Arctic more entrepreneurial thanks to immigration. In this aspect, the chapter contributes to existing studies on the linkage between immigrant entrepreneurship and socio-economic integration of immigrants by incorporating the concept of mixed embeddedness into a more exotic and underexplored process of immigration to the Circumpolar North.

The chapter first presents the motivations for this exercise, showing the population challenges of the Arctic with a special emphasis on immigrants and their entrepreneurial activities. Then, it describes the main assumptions of the mixed embeddedness theory and its most recent extensions and developments. Finally, some case studies are presented that link the successful integration of immigrant entrepreneurs to the concept of community resilience and discuss the applicability of these studies to the Arctic context. The chapter ends with suggestions for future research on immigrant entrepreneurship and recommendations for policymakers in the Arctic.

Population challenges in the Arctic: the role of immigration and migrant entrepreneurship

The Arctic is one of the most sparsely populated areas of the world. The small population of the region can be explained by the harsh climate and few economic opportunities. Moreover, after the collapse of the Soviet Union and the end of the Cold War, most of this region has undergone negative demographic processes (Khoreva et al., 2018). The emblematic example to depict the depopulation process in this period is the case of Pyramiden, once a Soviet coal-mining city in the Svalbard archipelago with more than 1000 inhabitants, which was abruptly abandoned in 1998. The whole Arctic has lost 2.4 million inhabitants in the last 30 years (1990–2018; see Figure 10.1).

The most affected areas in this regard were the regions of the Russian Arctic (a net population loss of 1.1 million), but more developed Finland and Sweden also experienced significant reductions in their Arctic populations, whereas Greenland managed to maintain a stable population.

Moreover, with the notable exceptions of Iceland and the Canadian Arctic provinces, much of this population decline could be attributed to negative migration balance: both internal (within an Arctic country) and international (Figure 10.2). This out-migration is not only driven by the lack of economic opportunities, but also by climate change: owing to the thawing permafrost, some local communities are endangered by land erosion or flooding and forced to move (Hamilton et al., 2016). This means that, despite the rapid development

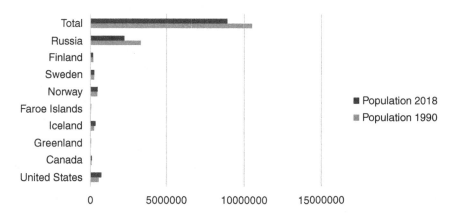

Figure 10.1 Population of the Arctic by country (1990–2018)

Source: Heleniak et al. (2019)

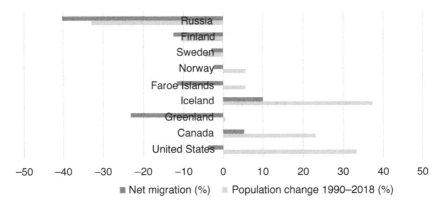

Figure 10.2 Population change and net migration in the Arctic by countries (1990–2018)

Source: Heleniak et al. (2019)

of transport, communication and construction technologies in the last 30 years, the Arctic regions have been unable to reverse the negative demographic trends. As most of the Arctic regions have much older age structures and lower fertility rates than the World average (Heleniak & Bogoyavlensky, 2015),[3] the main potential channel for population growth is to attract both internal and international migrants.

In spite of this gloomy picture, the region is perceived by most of the Arctic countries as strategic, and prospects for rapid future economic development look real. The climate change that has such a negative impact on indigenous communities of the Arctic at the same time offers huge possibilities for economic growth. For instance, with a milder climate in the region, the attractiveness of the Northern Sea Route (NSR) and the Northwest Passage (NWP) linking Western Europe and North America with Asia is increasing. The NSR, which includes the northern Eurasian coast, when free of ice (which currently happens for about 1.5 months per year), shortens the distance for maritime transport from northern Europe to north-east Asia and north-west North America by about 50 per cent compared with the southern route through the Suez or Panama Canal. Additionally, the NWP, which follows the northern North American coast, shortens maritime transport from Western Europe to the Far East in comparison with the Panama Canal route (Khon et al., 2010). If these routes become operational in the nearest decades, the coastal Arctic cities along them will automatically gain political, but also economic, importance. Also, the availability of Arctic resources will increase owing to climate change, attracting large-scale investments in the hydroelectric, oil, gas and mining industries (Prowse et al., 2009). However, the main challenge for the Arctic subregions in the near future is to develop more heterogeneous and sustainable strategies for future development: instead of focusing only on transport infrastructure and mining industries, this means also promoting investment in tourism and other, maybe more innovative, sectors of the economy.

In this regard, international migration to the Arctic could become the key driver of economic change. There are at least three main channels through which immigrants can benefit the Arctic economies. First, they enhance the entrepreneurial spirit of local communities, contribute to knowledge spillover between ethnic communities and the native population and, through their social contacts with their home countries, could ease the internationalization of newly established businesses. It is well known that the inflow of diverse ethnic groups usually enhances the creativity of the receiving society. Second, they might have a more open and creative vision of economic matters compared with indigenous actors, which enables them to perceive and capture economic opportunities. In doing so, they can be more effective as managers of local enterprises, contributing to their faster expansion. Finally, immigrants usually exhibit much higher entrepreneurial rates than natives. Therefore, as in the case of other regions of immigration, they could boost business creation in the Arctic (Yeasmin, 2016), contributing to job creation not only for themselves and their ethnic counterparts, but also for the natives (Brzozowski & Lasek, 2019). Moreover, immigrant entrepreneurs can positively affect trade volumes and even revitalize certain sectors or areas (Munkejord, 2017; Barberis & Solano, 2018). In the case of the Arctic, the entrepreneurial rates and investments in innovative activities are quite low: for instance, the Arctic areas of Norway, Sweden and Finland have 0.14 patent applications to the national office per 1,000 capita compared with national averages of 0.22 (average rates for the period 2003–2015). The business density rates are not impressive either: as of 2012, the Faroe Islands has about one company per 21 people, and Greenland fares a bit better (one company per 14 people in 2015), and other Nordic states' rates are relatively low (from 13 to 24 people per company; cf. Nordic Council of Ministers, 2018). Thus, enabling an inflow of entrepreneurial foreigners into the Arctic could mean a complete "game changer": if managed properly, international migration could become one of the main driving forces of the economic development for future decades. Yet, the main challenge for the Arctic regions is to build community resilience: a system that would enable successful insertion of immigrants into a new destination and would facilitate the creation and development of immigrant enterprises. In this aspect, the theory of mixed embeddedness could serve as a good starting point to revise the existing policies and activities directed towards integration of immigrants and their entrepreneurial activities.

Mixed embeddedness concept and its further developments

The original mixed embeddedness thesis was built as a critical response to the neoclassical model of entrepreneurial opportunity. According to the neoclassical paradigm, the opportunity structure, understood as market conditions and access to ownership (including degree of competition, business vacancies and public policies regarding entrepreneurship; cf. Waldinger et al., 1990), is fully transparent and equally accessible to both indigenous and immigrant individuals. As this framework obviously does not match the reality, especially in the case of immigrants from less-developed countries who settle in more advanced economies,

Kloosterman and Rath (2001) called for closer attention to be paid to the demand side of entrepreneurship and to the barriers and constraints faced by immigrants who consider creating a business in the host country. Consequently, they proposed the concept of mixed embeddedness: in order to understand the socio-economic insertion of immigrant entrepreneurs, it is necessary to consider not only their embeddedness in co-ethnic social networks, but also their embeddedness in their socio-economic and politico-institutional environments in the destination (Kloosterman & Rath, 2001).

When discussing the opportunity structure, Kloosterman and Rath (2001) emphasize two aspects that are crucial for successful integration of immigrant entrepreneurs: the accessibility of markets for newcomers willing to open a business and the growth potential of these markets. These two factors, crucial for immigrant entrepreneurship development, should be analysed at national, regional and local levels. For instance, at national level, there might be formal regulations that constrain the creation of immigrant entrepreneurs, imposing bureaucratic, legal or financial conditions that are very hard to meet for recently arrived foreigners. The regional level affects immigrant entrepreneurship, as the regions are very heterogeneous and tend to specialize in different economic activities. In this context, the relative peripherality of the Artic subregions and their specialization in coal mining, oil and gas extraction provide few opportunities for prospective immigrant entrepreneurs, who usually do not have much capital and create businesses mostly in the SME sector. Also, the local context is important: there are towns and neighbourhoods where immigrants are highly concentrated and can rely on the support of the ethnic enclave, and there are locations where immigrant communities are too weak for them to do so (Zubair & Brzozowski, 2018). On the other hand, there might be negative factors at play also at the local level. These are the policies that exclude migrants by giving preference to traditional local industries: for instance, in the case of some small towns in Lombardy (Italy), kebab shops were banned to protect traditional Italian trattorias (Barberis & Solano, 2018).

Further elaborations of the mixed embeddedness model point to the "dual nature" of immigrant insertion in the host country: (1) embeddedness in the socio-institutional structure in which their business is started and then developed (meso level of opportunity structure and macro institutional framework) and (2) embeddedness in the immigrants' networks of social relations (and immigrants' resources; cf. Kloosterman, 2010). This enables identification of conditions that are favourable to immigrant entrepreneurship: the markets in a host country have to be open to the new entrants, not only in terms of economic barriers, but also when it comes to national and local rules and regulations. Based on this theoretical framework, a typology of opportunity structure was developed, enabling comparisons of conditions of entry into entrepreneurship and its future growth prospects (Figure 10.3).

The upper-left area of Figure 10.3 remains blank, as sectors that are stagnating and require high human capital levels are unattractive for immigrants who either seek entrepreneurial activity in expanding sectors or simply consider waged employment. The vacancy-chain openings are the opportunity structures

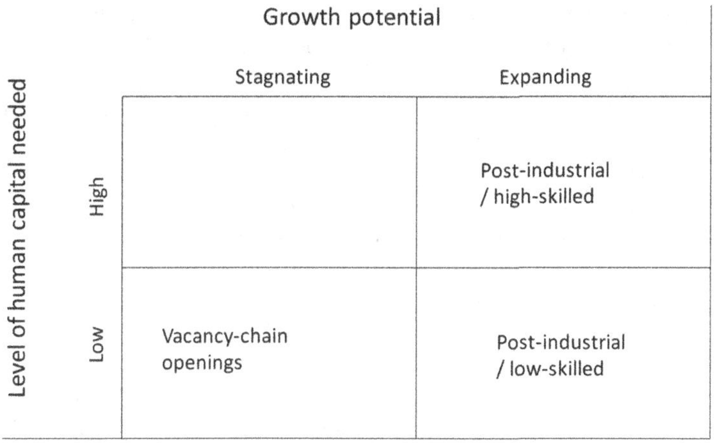

Growth potential

Figure 10.3 Typology of opportunity structure in the mixed embeddedness model

Source: Kloosterman (2010)

characterized by low entry requirements and, therefore, they are very often considered as an attractive option for immigrant entrepreneurs. Unfortunately, such markets are usually stagnating or even in decline: these are very often sectors left over by natives in which the profits are low and the risks of failure are high. Very often, survival in this market is possible only in the case of reliance on low-paid work by family members and in the ethnic markets dominated by members of the same ethnic community. Typical examples of such businesses include Asian liquor stores in the US, trading points for clothes or shoes on open-air fair trades in Central and Eastern Europe, Indian internet kiosks in Southern and Western Europe and Arab/Turkish kebab shops in most parts of Europe.

The post-industrial, low-skilled markets are characterized by relatively low entry barriers and high growth potential, and so, compared with vacancy-chain openings, they are much more attractive alternatives for immigrant entrepreneurs. These activities mostly comprise services, such as household and care services for which there is continuously increasing demand in North America and North-west Europe. Finally, in the top-right corner, there are markets that require high levels of human capital, but, on the other hand, they represent a high growth potential. These are usually innovative, high-technology firms. Although these businesses are often perceived as elite (for instance, ICT), immigrants are also increasingly successful in these sectors (Brzozowski et al., 2014).

As most of the immigrant entrepreneurs are usually clustered in the vacancy-chain opening opportunity structure, the chances of their economic success are limited. Therefore, Kloosterman suggests that the most suitable way for the immigrant entrepreneur and their family to achieve upward mobility is a breaking-out strategy – that is, a shift of a business model into more profitable sectors. In this regard, the most likely transition is that into the post-industrial/low-skilled market: it requires some experience which could be gained by an immigrant during the initial stages of entrepreneurial activity, supplemented by a change in production and adjustment of a business model (Kloosterman, 2010). However, the crucial factor needed to succeed in breaking out is the widening of the immigrant's social capital and, more precisely, wider access to weak ties (cf. Granovetter, 1985) and social contacts beyond the co-ethnic community and homogeneous ethnic networks. There are multiple examples of such successful immigrant enterprises operating in post-industrial/low-skilled markets, which include such services as delivery and packaging, catering and food, repair and maintenance (Kloosterman & Rath, 2018), but also ethnic-specific ones such as Ayurveda medicine, acupuncture sessions or yoga training.

In a recent special issue devoted to contemporary applications of mixed embeddedness theory, Barberis and Solano (2018) show that this concept has influenced at least three areas of migration studies: the spatial dimension of migrants' embeddedness, the super-diversity debate and research on immigrants' transnationalism. The first strand of research is particularly important: it reminds researchers not only to consider the three-level approach developed by Kloosterman and Rath (2001) – national, regional/urban and neighbourhood levels – but also to investigate the role that immigrant entrepreneurs play in restructuring local economies in which they settle.

In an essay that summarizes 20 years of mixed embeddedness theory, Kloosterman and Rath (2018) point out its most important extension: the dynamism of immigrant entrepreneurship. The typology and model of opportunities outlined by Kloosterman (2010) is not static, as it enables movement from the vacancy-chain opportunities into more promising post-industrial markets, as the migrants accumulate strategic resources. Moreover, the institutional framework can change over time, giving more market opportunities to immigrant entrepreneurs, but also sometimes closing some of them. For instance, the enforcement of employment regulations in Amsterdam's garment industry has severely hit Turkish entrepreneurs, as many of them relied on illegal agreements (Kloosterman & Rath, 2018).

Mixed embeddedness in practice: linking successful integration within community resilience

In her study of immigrant entrepreneurship in Lapland, Yeasmin (2016) emphasizes that a substantial number of businesses created in this region by immigrants are necessity-driven. As the unemployment rate for foreigners is more than double the overall unemployment rate in Lapland, foreigners are forced to become self-employed, looking for any possibilities to sustain themselves and their families. In this sense, most of the immigrants who run business activities are – according to Kloosterman's model (2010) – trapped in vacancy-chain openings, with very limited chances to develop their business and make it economically sustainable. When investigating the opportunity structure of the region, Yeasmin emphasizes that, in principle, the institutional constraints on opening a business are limited: in fact, it is very easy to start a new firm. Yet, it seems very difficult to keep this enterprise running for a longer period and make it profitable. As most of the immigrant businesses in Lapland are typical ethnic firms whose produce or services are directly linked to immigrants' culture – that is, ethnic restaurants, ethnic medicine services and so on – the main obstacle to further growth of such businesses is the fact that, "the market of ethnic consumers is small and demand is not sufficient for running a business dependent solely on ethnic consumption" (Yeasmin, 2016: 134).

Therefore, it seems that the main constraint on immigrant entrepreneurs in this case is the absence of an ethnic enclave economy. The ethnic economy is a concept developed by Light and associates (1994) and defined as a subsector of the host country economy dominated by migrant/minority entrepreneurs and the ethnic employment sector. Then, the ethnic enclave is a part of ethnic economy in which there is a substantial number of immigrant firms and employees from the same ethnic group. This specific geographical ethnic concentration brings several benefits for immigrant entrepreneurs: they can take advantage of vertical and horizontal integration. This means that they are linked and can cooperate with co-ethnic suppliers and customers, rely on informal ethnic institutions that provide access to financial capital and know-how, and can employ co-ethnic workers who are more flexible, less demanding and ready to work longer and harder than the indigenous employees. In this case, the entry into self-employment is relatively easy, but only for the individuals who come from a specific ethnic group, and it is closed to competition for outsiders. Although the ethnic enclave economy is regarded as providing a fertile ground for immigrant business creation, at the same time, many authors indicate that it is very difficult for such entrepreneurs to further develop their business and make it sustainable in the long run – that is, providing a stable source of income for the migrant and their family (Zubair & Brzozowski, 2018).

Yet, in the Arctic case, access to such a specific "entrepreneurial incubator" as the ethnic enclave is not possible owing to the small size of immigrant populations and low ethnic concentration (Yeasmin, 2016). In such a case, it is interesting to return to the examples of immigrant entrepreneurs who managed to succeed without the support of the ethnic enclave. In doing so, we rely on the study by Zubair and Brzozowski (2018), who investigated immigrant entrepreneurs in two European cities: Klagenfurt in Austria and Krakow in Poland. At the time the study was carried out (years 2014–2016), the immigrant populations in both cities were relatively small, and so immigrant entrepreneurs could not rely on ethnic enclave links and informal institutions when conducting their business activities. Most of these businesses were in fact created with the help of family networks and resources, and not thanks to co-ethnic resources. Those immigrants who tried to develop the firm in a "traditional" way, maintaining operations in vacancy-chain openings and trying to stick with co-ethnic customers, partners or suppliers usually failed or were thinking of closing their businesses at the time of the survey (Zubair & Brzozowski, 2018).

Yet, an interesting picture arises when we look at those immigrants who took a more innovative stance and decided to look for business contacts and clients beyond their close, homogeneous co-ethnic network. An obvious choice in Klagenfurt was for a Sikh entrepreneur who was running a grocery store to approach other immigrants from Asia, including Indians, Afghanis, Pakistanis or Thais. Moreover, this grocery shop evolved over time, also offering, apart from food products typical of Asian cuisine, electronics, telecommunication and money transfer brokerage services and even tickets for cultural events, such as the Bollywood movie festival in a local cinema. Sometimes, however, this diversification of business activity was not enough to make the business sustainable, and economic survival depended on overcoming the traditional ethnic barriers and conflict. For instance, a very patriotic Palestinian entrepreneur was able to develop his restaurant by relying on close cooperation with a Jewish hotel owner who marketed the business among his Jewish clientele. Finally, the most promising option was a break-out strategy, reaching to the mainstream, post-industrial, low-skilled market. This was the case of a Pakistani entrepreneur who ran a bar and café. Being successful in this business was not easy and required not only discarding traditional Islam values (as most of the revenue came from selling alcoholic drinks), but also operating during non-traditional opening hours – actually, his business was open exactly during the times when most rival bars were closed (Zubair & Brzozowski, 2018).

The study by Munkejord (2017) is even more relevant, as it deals with the network embeddedness of immigrant entrepreneurs in the context of Finnmark and the Arctic region of Norway. The most novel aspect of this research was that it demonstrates that recently arrived immigrants are able to successfully create and exploit entrepreneurial opportunities. Sometimes, this rapidly achieved embeddedness in local networks is possible thanks to locals' openness to newcomers, as in the case of the immigrant from Switzerland who started to produce handcrafted glass ornaments with graphic elements inspired by Nordic culture and landscape. The local community was so interested in these novel products that one of the local entrepreneurs quickly offered the manufacturer the possibility of selling the ornaments at a local petrol station. Other members of the local community started to buy the ornaments, feeling a sense of regional pride in the fact that the ornaments are made locally and the designs are inspired by local folk tradition.

Yet, when the cultural distance between the immigrant and the native community is substantial, becoming embedded in local networks takes more time. This was the case for an African refugee who, just a few weeks after arrival, created an association for refugees and a local gospel choir and a band. His music activities served as a way to extend his personal network, both with other immigrants and with the representatives of the indigenous population. He also took advantage of the introductory programme for refugees to learn Norwegian. After acquiring enough cultural and social capital, he was then able to open a small café and a shop that served both immigrants and the local population and became an important venue where newcomers could socialize with local community members (Munkejord, 2017). The first case, the handicraft production of glass ornaments, shows not only the significant innovativeness and flexibility of the immigrant entrepreneur involved, but also the relatively high level of local community resilience and readiness to quickly integrate newcomers. The second case, the refugee entrepreneurship, also shows the local community's good resilience mechanisms, inclusiveness and the availability of institutions (the local church in which the gospel choir and band played and local authorities that helped to create a refugee association).

Conclusions

The case studies of successful immigrant entrepreneurs presented in this chapter could serve as examples of good practice for the establishment of community resilience in several Arctic subregions. The lessons stemming from the mixed embeddedness model are quite simple: it is important to create incentives for immigrant entrepreneurs to adopt break-out strategies from vacancy-chain openings to more profitable market sectors that offer changes to a firm's development. As the immigrant communities in the Arctic are small and usually deconcentrated, they lack the support of the ethnic enclave. Thus, it is crucial to provide sound support structures for immigrants willing to engage in entrepreneurial activities. It is very important to allow for a gradual insertion and embeddedness into the local community, without pressure to create a business as soon as possible. This is especially likely in the case of refugees and other politically driven migrants. Such support structures could be social enterprises, including community-based social enterprises, that would function as a "preschool" for future migrant entrepreneurs who lack experience and resources and are not yet legally entitled to open a business (for instance, asylum applicants). These social enterprises could work as potential hubs in which shared working space could be combined with training events, but they could also provide counselling or tutoring assistance in drafting a business plan for a future business. It is equally important to involve the representatives of the local community in such activities – as either tutors, clients of the social enterprise or even co-workers. In such a case, immigrants broaden their social networks, which are then very useful for their firms.

Another important avenue for entrepreneurship development and the expansion of immigrant entrepreneurship would be public support and incentives for transnational immigrant businesses (Brzozowski et al., 2014) – that is, businesses that are run by immigrants simultaneously in the host (i.e. the Circumpolar North) and home countries. Such enterprises could serve as natural economic bridges, linking the Arctic with the developing world and enabling the expansion of Arctic exports to these countries.

Notes

1 For instance, since 1989, the Russian North has lost ca. 17% of its population owing to outmigration, and, in the extreme case of Chukotka, three-quarters of the population have left (Heleniak, 2012).
2 The author is perfectly aware of the complexity of definitions of immigrant entrepreneurship, but, for the purpose of this study, it is simply understood as entrepreneurial actions undertaken by foreign-born persons.
3 The only exceptions in this regard are the indigenous populations, who are at the earlier stages of demographic transition.

References

Barberis, E., & Solano, G. (2018). Mixed embeddedness and migrant entrepreneurship: hints on past and future directions. An Introduction. *Sociologica*, 12(2), 1–22.

Berkes, F., & Ross, H. (2013). Community resilience: toward an integrated approach. *Society & Natural Resources*, 26(1), 5–20.

Brooks, M. R., & Frost, J. D. (2012) Providing freight services to remote Arctic communities: are there lessons for practitioners from services to Greenland and Canada's northeast? *Research in Transportation Business & Management*, 4, 69–78.

Brzozowski, J., Cucculelli, M., & Surdej, A. (2014). Transnational ties and performance of immigrant entrepreneurs: the role of home-country conditions. *Entrepreneurship & Regional Development*, 26(7–8), 546–573.

Brzozowski, J., & Lasek, A. (2019). The impact of self-employment on the economic integration of immigrants: evidence from Germany. *Journal of Entrepreneurship, Management and Innovation*, 15(2), 11–28.

Duhaime, G., & Caron, A. (2006). The economy of the circumpolar Arctic. In S. Glomsrød & I. Aslaksen (Eds.), *The economy of the North* (pp. 17–23). Oslo: Statistics Norway.

Exner-Pirot, H. (2015). Innovation in the Arctic: squaring the circle. Paper presented at Arctic Summer College. Retrieved on August, 8, 2019 at: www.arcticsummercollege. org/sites/default/files/ASC%20Paper_Exner-Pirot_Heather_0.pdf

Granovetter, M. (1985). Economic action and social structure: The problem of embeddedness. *American Journal of Sociology*, 91(3), 481–510.

Hamilton, L. C., & Mitiguy, A. M. (2009). Visualizing population dynamics of Alaska's Arctic communities. *Arctic*, 62(4), 393–398.

Hamilton, L. C., Saito, K., Loring, P. A., Lammers, R. B., & Huntington, H. P. (2016). Climigration? Population and climate change in Arctic Alaska. *Population and Environment*, 38(2), 115–133.

Heleniak, T. (2012) International comparisons of population mobility in Russia. *International Journal of Population Research*, 2012, 1–13.

Heleniak, T., & Bogoyavlensky, D. (2015). Arctic populations and migration. In J. N. Larsen & G. Fondahl (Eds.), *Arctic human development report: regional processes and global linkages* (pp. 53–104). Copenhagen: Nordisk Ministerråd.

Heleniak, T., Turunen, E., & Wang, S., (2019). Cities on ice: population change in the Arctic. Retrieved on August 8, 2019 at: www.nordregio.org/nordregio-magazine/ issues/arctic-changes-and-challenges/cities-on-ice-population-change-in-the-arcti

Khon, V. C., Mokhov, I. I., Latif, M., Semenov, V. A., & Park, W. (2010). Perspectives of Northern Sea Route and Northwest Passage in the twenty-first century. *Climatic Change*, 100(3–4), 757–768.

Khoreva, O. G., Konchakov, R., Leonard, C. S., Tamitskiy, A., & Zaikov, K. (2018). Attracting skilled labour to the North: migration loss and policy implications across Russia's diverse Arctic regions. *Polar Record*, 54(5–6), 324–338.

Kimmel, M., Farrell, C. R., & Ackerman, M. (2019). Newcomers to ancestral lands: immigrant pathways in Anchorage, Alaska. In S. Uusiautti & Y. Nafisa (Eds.), *Human migration in the Arctic* (pp. 93–116). Singapore: Palgrave Macmillan.

Kloosterman, R., & Rath, J. (2001). Immigrant entrepreneurs in advanced economies: mixed embeddedness further explored. *Journal of Ethnic and Migration Studies*, 27(2), 189–201.

Kloosterman, R. C. (2010). Matching opportunities with resources: a framework for analysing (migrant) entrepreneurship from a mixed embeddedness perspective. *Entrepreneurship and Regional Development*, 22(1), 25–45.

Kloosterman, R. C., & Rath, J. (2018). Mixed embeddedness revisited: a conclusion to the symposium. *Sociologica*, 12(2), 103–114.

Krasulina, O. Y. (2018). Problems of entrepreneurship development in the Russian Arctic Zone. *IOP Conference Series: Earth and Environmental Science*, 180(1), 1–6.

Light, I., Sabagh, G., Bozorgmehr, M., & Der-Martirosian, C. (1994). Beyond the ethnic enclave economy. *Social Problems*, 41(1), 65–80.

Merhar, A. (2020). Embodying transience: Indigenous former youth in care and residential instability in Yukon, Canada. In N. Yeasmin, W. Hasanat, J. Brzozowski & S. Kirchner (Eds.), *Immigration in the circumpolar North: Integration & resilience* (pp. 00–00). Abingdon, UK, and New York: Routledge.

Munkejord, M. C. (2017). Becoming spatially embedded: findings from a study on rural immigrant entrepreneurship in Norway. *Entrepreneurial Business and Economics Review*, 5(1), 111–130.

Nilsen, T. (2016). Why Arctic policies matter: The role of exogenous actions in oil and gas industry development in the Norwegian High North. *Energy Research & Social Science*, 16, 45–53.

Nordic Council of Ministers. (2018). *Arctic business analysis. Entrepreneurship and innovation*. Copenhagen: Nordic Council of Ministers.

Parfenova, E., Tchebakova, N., & Soja, A. (2019). Assessing landscape potential for human sustainability and "attractiveness" across Asian Russia in a warmer 21st century. *Environmental Research Letters*, 14(6), 065004.

Prowse, T. D., Furgal, C., Chouinard, R., Melling, H., Milburn, D., & Smith, S. L. (2009). Implications of climate change for economic development in northern Canada: energy, resource, and transportation sectors. *AMBIO: A Journal of the Human Environment*, 38(5), 272–282.

Romashkina, G. F., Didenko, N. I., & Skripnuk, D. F. (2017). Socioeconomic modernization of Russia and its Arctic regions. *Studies on Russian Economic Development*, 28(1), 22–30.

Waldinger, R., Aldrich, H., & Ward, R. (1990). Opportunities, group characteristics and strategies. In R. Waldinger, H. Aldrich & R. Ward (Eds.), *Ethnic entrepreneurs: immigrant business in industrial societies* (pp. 13–48). London: Sage.

Yeasmin, N. (2016). The determinants of sustainable entrepreneurship of immigrants in Lapland: an analysis of theoretical factors. *Entrepreneurial Business and Economics Review*, 4(1), 129–159.

Ziaja, W., Dudek, J., Ostafin, K., Węgrzyn, M., Lisowska, M., Olech, M., & Osyczka, P. (2016). Environmental and landscape changes. In W. Ziaja (Ed.), *Transformation of the natural environment in Western Sørkapp Land (Spitsbergen) since the 1980s* (pp. 35–49). Cham, Switzerland: Springer International.

Zubair, M., & Brzozowski, J. (2018). Entrepreneurs from recent migrant communities and their business sustainability. *Sociologica*, 12(2), 57–72.

11 Migration and sustainable development in the European Arctic

Stefan Kirchner

Introduction

The circumpolar Arctic is an area of great diversity but shared challenges. Hardly any other part of the Arctic is as accessible as the continental European High North (EHN), the Arctic and subarctic region consisting of the northernmost parts of Norway, Sweden and Finland, comprising the Norwegian counties of Nordland, Tromsø and Finnmark, the Swedish counties of Norrbotten and Västerbotten, as well as the Finnish region of Lapland. Climate change and globalization are the two key issues affecting the Arctic today. The same applies to the EHN, which has a population of approximately 1.2 million people (including tens of thousands of indigenous Sámi) but, at about 446,000 km², is larger than Japan or Germany and about the size of Sweden. In recent years, like other parts of Europe, this sparsely populated area has seen an increase in the number of migrants and, in particular, refugees. This has proven to be both a challenge and an opportunity for small communities in the far north. Mobility is a key element of our globalizing world (see e.g. Bederman, 2008: p. 55 et seq.) and it also affects the Arctic, most notably in the forms of the mobility of money and investments and the mobility of people –that is, tourism and migration.

Although an attempt to draw a complete picture of the situation across the entire EHN would be beyond the scope and the page limitations of this chapter, it will attempt to show the role immigration can play in sustainable development in the region. Particular attention will be paid to the role rural communities can play in the integration of migrants, especially refugees, and how local communities in the EHN might benefit from increasing migration.

Climate change and globalization in the Arctic

According to the Swedish government's official Arctic policy,

> [t]he Arctic is in a process of far-reaching change. Climate change is creating new challenges, but also new opportunities, on which Sweden must take a position and exert an influence. New conditions are emerging for shipping, hunting, fishing, trade and energy extraction, and alongside this, new needs

are arising for an efficient infrastructure. New types of cross-border flows will develop. This will lead state and commercial actors to increase their presence, which will result in new relationships.

(Sweden, 2011, n.p.)

Nothing has changed the Arctic and the living conditions of the people in the far north as rapidly as climate change, which is the fundamental challenge of our time, and globalization, the fundamental opportunity gained by humanity in recent years. Despite the current debate about trade wars and limits to free trade, the essential direction of the development in this regard remains unchanged, and globalization is more than a trend: it is a key characteristic not only of the global economy, but of human society as a whole. Unfortunately, this economic freedom is not yet reaching everybody in the Arctic (as is the case in many other parts of the world as well). At the moment in history when the world looks at the Arctic both with concern and optimism, the economies of many northern communities are limited, often to one or just a few forms of income, such as extractive industries or tourism. Although these income streams are facilitated by globalization, they are in themselves insufficient for the creation of truly sustainable local industries. Climate change, meanwhile, puts additional pressures, in particular, on traditional and indigenous livelihoods in the Arctic owing to the radical changes in the natural environment that is essential for the successful conduct of, for example, reindeer herding.

Economic development in the European High North today: extractive industries and tourism

Despite the low population density and the distance from urban centres where decisions are often made, the Arctic is not an empty space but home to millions of people, some of them newcomers like myself, others tracing their families' histories in this land for thousands of years. This is hardly a new phenomenon. For centuries, the Arctic has been a destination for those seeking new opportunities. In his bestseller *Into the Wild*, Jon Krakauer (2007: 97) cites Fridtjof Nansen, who mentioned Irish monks who reached a small island off the southern coast of Iceland in the search for tranquillity as early as in the 5th or 6th century A.D. Interestingly, a similar voyage was reported for the year 458 A.D., when Buddhist monks are said to have sailed from China, via the Kuril Islands, Kamchatka and the Aleutian Islands, to what is today Alaska (Lopez, 2014: 339). The Arctic has long been a space for migratory movements (see Durfee & Johnstone, 2019: 35 et seq.), but, for a long time, though, it has only been the indigenous peoples of the Arctic who have truly made the Arctic their home. Although recent centuries have seen a dominance of settler societies, there is still an imbalance between centres and peripheries in the North. This can be seen across the Arctic, as all Arctic areas are governed by states that have their economic and political power centres far from the Arctic.

That rural communities can also be disadvantaged in highly developed countries might be illustrated by an example at the crossroads of economic development and environmental protection: one of the problems concerning the protection of the natural environment in the Arctic can be traced back to the origins of the environmental movement, which often was seen as disconnected from rural communities. Writing about the situation in the United States, Alice Outwater found that,

> [w]hen the environmental movement began, it appealed principally to well-educated, well-heeled white people. Environmental concerns correlated with non-rural residence, were strongest among those fifty-five and older, and arose independent of political affiliation. At the time, it was said that environmentalists were people who had already bought their second homes: these people had something worth protecting, and they were old enough to remember a time when the waters weren't polluted and the air was clean. They also had access to power and understood the workings of the political process.
>
> (Outwater, 1996: 162)

In practice, however, dependence on an intact natural environment is even more important for rural residents, especially for indigenous peoples, but they frequently lacked access to power and often also access to information. The realization of the threat posed to nature often reached rural communities very late. This is not only the case in the Arctic but also elsewhere. Outwater's description of the situation in the United States at the beginning of the era of environmental consciousness reflects experiences elsewhere, too. The imbalance of power, be it subjective or objective, is still felt acutely in Arctic regions, which are usually the peripheries of their respective countries. Combined with the colonial history of the EHN and the relative lack of political importance of sparsely populated Arctic and subarctic parts of Norway, Sweden and Finland when compared with the population centres in the South, this power imbalance contributed to a situation in which the natural environment of the EHN is now at risk from the effects of economic development. It is not entirely without irony that the emphasis on equality and aversion to the idea of "special" treatment for specific groups, which is common in the Nordic countries, may actually hinder solutions to this problem.

Different states in the Arctic have chosen different approaches to economic development (on the different economic strategies of Arctic states, see Durfee & Johnstone, 2019: 138 et seq.), some of which are more sustainable than others. Whereas Norway is notable for its long-term plans to move away from over-reliance on hydrocarbon extraction, and the Sámi home area in Finland has so far only seen limited mining (Kirchner, 2018), Sweden continues to rely heavily on mining. Similarly, hydrocarbon extraction in the Arctic is a significant part of long-term economic planning in the Russian Federation. Hydrocarbons are of crucial importance for Russia (Marshall, 2016: 13), and the lack of permanently ice-free ports is a key issue for Russia (cf. Marshall, 2016: 31). This might make it easier to understand why, from Moscow's perspective, climate change is not seen as a problem but as an opportunity – for example, with the opening of the Northern Sea Route (Marshall, 2016: 273). This view is shared by companies that build their business on exploiting the natural resources of the Arctic (cf. Marshall, 2016: 274, 280 et seq.). Although some argue in favour of using the riches of the Arctic for all (Marshall, 2016: 282), the people who live in the Arctic and will eventually pay the price for increasing hydrocarbon extraction activities must not be forgotten. After all, things look very different not only in Russian towns and villages that are sinking into the ground as the permafrost melts, but also from the perspective of other Arctic countries. All of these problems are wake-up calls for local communities in the Arctic, and there is already increasing interest in sustainable development solutions for the Arctic. Unfortunately, there appears to be a lack of interest in, let alone concern for, the Arctic, its people and its nature among many of those who come to the Arctic specifically for the purpose of hydrocarbon extraction (Lopez, 2014: 397). The Arctic tourism boom, meanwhile, is evidence that there is a fundamental human desire for nature (Hilton, 2016: 359). It is this need to connect with nature that has given rise to the environmental movement, the notion of environmental human rights and increasing recognition of indigenous environmental rights (see e.g. Westra, 2008: 3 et seq. and 219 et seq.).Today, connecting with nature can be considered a human right (Heinämäki, 2010). However, despite the growth of green industries, so far, sustainable development in the Arctic is more aspiration than reality.

Right now, the Arctic is undergoing massive changes, the causes of which are to be found, to a large degree, outside the Arctic. Although globalization and technological development provide opportunities for the people of the Arctic, climate change is an unprecedented challenge for the Arctic. The Arctic region is warming twice as fast as the global average, and the effects of climate change can already be felt today. In recent winters, the tourism sector here has suffered significantly owing to a lack of snow in November and early December, the prime tourism season, and reindeer herders have been affected by changing snow and vegetation patterns for years, to give just two examples. As climate change makes the Arctic more accessible, indirect effects, such as water pollution from mining or infrastructure projects, are also likely to become more of a problem in the future.

Tourism is, at times, seen as an alternative form of income, and a number of locations in the EHN are placing a lot of faith in the economic potential of tourism. That there is currently a boom in Arctic tourism is undeniable. It remains to be seen if this is a short-term boom or a long-term trend. The characteristics of so-called "overtourism" are already becoming visible in a number of locations – for example, in Rovaniemi, the capital of the Finnish region of Lapland. Overtourism is affecting daily life in places as diverse as Lisbon, Barcelona, Amsterdam and Krakow – relatively easily accessible cities with a reputation for being more affordable than larger cities such as London, Paris or New York. In the EHN, tourism is often limited to the winter months (although there are some efforts to increase the number of visitors for the summer). Because the limited infrastructure in the EHN is often insufficient to handle a large number of visitors, some places are experiencing a boom in the construction of hotels and winter sport infrastructure, which may remain largely unused for significant parts of the year. This boom benefits not necessarily local residents, but, for example, international investors. At the same time there is the increasing demand on housing – for example, through nominally share-economy rental schemes – and transportation, especially between periphery towns and larger cities such as Helsinki, caused by larger tourist numbers and driving up costs for local residents as well. Although less destructive than the extractive industries at first sight, tourism does not provide a guarantee for sustainable development in the EHN either.

Like many developments in the Arctic, infrastructure is a double-edged sword. On the one hand, for example, the construction of roads means better connectivity and improved access, for instance, to health care or education services; on the other hand, roads and especially railways pose a significant risk for reindeer, which are an important source of income, especially for indigenous Arctic communities, but also of great cultural relevance. When it comes to developing Arctic economies, non-economic factors such as indigenous cultures have to be taken into account as well. In order to be effective and beneficial for people, sustainable development needs a holistic approach, including traditional livelihoods as well as high tech.

In order for the economies of the Arctic to be developed in a manner that is truly sustainable and does not deplete natural resources, the people who live and work in the Arctic will require more options to make a living. These livelihood options can be old or brand new – they might even involve work we can hardly imagine today – but they will all require knowledge and education. At this time, job opportunities in many Arctic communities are limited to the extraction of non-living natural resources or tourism and to common forms of employment such as can be found elsewhere, in the service industries, public services and sales of products, and traditional livelihoods, such as fishing or reindeer herding. Such, often more sustainable, traditional sources of income are often made impossible owing to other economic activities, such as mining. Land use conflicts are a common theme in the Arctic and, in many places, are a constant source of conflict. This situation is worsened by climate change. This is not only because of the direct effects of climate change, but also because climate change makes many parts of the Arctic more accessible than ever before. This increased accessibility leads to more resource extraction and infrastructure construction – for example, for the tourism industry or the transportation of raw materials – and often to more conflicts between different forms of livelihood.

Integration of migrants in rural communities

Within Europe, the nation-states have failed to agree on a distribution of refugees across the European Union (Schwan & Zobel, 2019a), resulting in a disproportionate economic burden on the southern countries, in particular Greece, Italy and Spain. Although the new Commission is planning reforms (Lehmann, 2019), it has to be noted that, while states have failed to take adequate action, a number of municipalities that have been suffering from declining population numbers have realized the long-term demographic and economic potential of migration. In a

European society that is ageing, rural regions and small municipalities are already competing for people, and the arrival of new persons who are willing to become part of a local society and who want to contribute to that society's economic development in the long run is an opportunity for rural and remote municipalities. It should not be ignored, however, that these gains for small communities in affluent countries often come at the cost of a brain drain affecting countries that are already suffering from war or disaster, such as Syria or Iraq. The focus on states, rather than municipalities, has not solved the refugee crisis (Schwan & Zobel, 2019a), and austerity policies that had been implemented to counteract the 2008 financial crisis have often hurt local economies (Schwan & Zobel, 2019a), especially in terms of infrastructure (Schwan & Zobel, 2019a). The latter is a process that is accelerated as people move away from villages to cities. But, although inbound migration can help small communities, integration demands resources: taking in refugees requires investments in municipalities by the state (Schwan & Zobel, 2019a) and the participation of different types of stakeholder in decision-making processes at the municipal level (Schwan & Zobel, 2019a). Through formalized participation procedures, such as advisory councils that would have a consultative function and would complement, rather than replace, the democratically elected local representative and executive institutions (Schwan & Zobel, 2019a), municipalities would be enabled to take different perspectives into account (Schwan & Zobel, 2019a). This in turn requires the existence of democratic structures, traditions and experiences at the local level.

In this context, the "local level" refers not only to municipalities, but to every settlement and village, even if a local community only consists of a few inhabitants. Participation has to be practical and realistically possible, which can be a particular challenge in sparsely populated areas in the EHN, where a municipal centre might be impractically far from outer parts of the municipality. From the perspective of public law, such an approach also requires strong guarantees about the ability of municipalities or regions to handle their own affairs based on ideas of subsidiarity and local self-determination. (On the situation in the Nordic countries, see Martínez Soria, 2007: 1018 et seq.) After all, municipalities are the place where many individuals experience the most direct contact with public authorities (Schwan & Zobel, 2019a; Karakurt, no year: 2), and municipalities are the institutions that provide many of the services most relevant for residents (Schwan & Zobel, 2019a). It is also at the local, municipal level that the integration of migrants actually happens (Schwan & Zobel, 2019a). That municipalities also want to be heard in the context of the integration of migrants can also be explained using the concept of subsidiarity (Schwan & Zobel, 2019a). Subsidiarity is a concept that has served the European Union well. In light of the particular geographical situation in the EHN, subsidiarity seems an obvious approach to the relationship between municipalities and states in the North as well. Subsidiarity goes both ways (Schwan & Zobel, 2019a) and requires mutual respect between different levels of public authorities (for an opposing view, favouring the role of the state instead of municipalities in the integration of migrants, see Heisterhagen, 2019).

Therefore, the integration of migrants has to happen on the local level. Whereas political and legal decisions regarding migration often happen in capital cities, the actual work of integration usually happens on the periphery, at the local level, involving local people. This has already been observed in countries outside the region that have had more experience with refugees and migrants than the remote communities of the EHN. For example, in Germany, which has seen the arrival of a large number of refugees and migrants in recent years, municipalities and regions are supported by the public availability of the results of research that has been undertaken by research institutions (see e.g. Deutsches Institut für Urbanistik, 2016; Aumüller & Bretl, 2008: 7 et seq.). Often, they also receive financial support from industry-related institutions (Bertelsmann Stiftung, 2018), non-governmental organizations, associations of municipalities (Deutscher Städtetag, 2016; Kommunale Gemeinschaftsstelle für Verwaltungs-management, 2017a, 2017b), governments (Der Beauftragte der Bundesregierung für Migration, Flüchtlinge und Integration, 2019; Hessische Staatskanzlei, 2018b) and other actors. Financial support is provided by the state for the support of volunteer work within the municipalities (Hessische Staatskanzlei, 2018a). For example, in the German state of Hesse (German: *Hessen*), municipalities and districts (*Landkreise*) receive up to €30,000 per year for projects that are aimed at integrating refugees (Hessische Staatskanzlei, 2018a), but the clear expectation on the part of the state is that the work itself is done by locals on a voluntary basis (Hessische Staatskanzlei, 2018a). The focus of the responsibility, therefore, now rests with the local community rather than with the state. These amounts might be relatively small, given the enormity of the task at hand for many municipalities, but such funding highlights the respect of the state for the necessary work at the local level (for an overview on the integration of refugees in municipalities in Germany, see Aumüller, 2018). Volunteers at the local level are essential for the successful integration of migrants (Hessische Staatskanzlei, 2018b: 70), but they need to receive training (Hessische Staatskanzlei, 2018b: 71) and require adequate support structures (Hessische Staatskanzlei, 2018b: 71). It is noteworthy that the state government had to emphasize that volunteers who are already giving of their time should have their costs reimbursed (Hessische Staatskanzlei, 2018b: 71). The integration of refugees requires a participatory approach (Hessische Staatskanzlei, 2018b: 71) that involves multiple actors (Hessische Staatskanzlei, 2018b: 71), including political decision-makers at the local level (Hessische Staatskanzlei, 2018b: 71). In Germany, which is hosting almost 1.8 million persons seeking protection (Landeszentrale für politische Bildung Baden-Württemberg, 2019) – about 2 per cent of the total population (Landeszentrale für politische Bildung Baden-Württemberg, 2019) – it has been concluded that municipalities have been quick to react (Ohliger et al., no year: 35) and have been able to build on existing structures (Ohliger et al., no year: 35). This infrastructure includes local civil society entities, especially local non-governmental organisations, such as associations (*Vereine*), but there is a significant lack of resources and competences at the local level (Ohliger et al., no year: 35). Therefore, it can be concluded that the active participation of local civil society groups plays a significant role in the successful integration of migrants into the host society, and that public authorities can directly strengthen civil society efforts in this regard through targeted funding.

This approach appears to be compatible with the political and societal situation in the European Arctic and subarctic regions as well. The EHN, at least the Nordic countries, provides an attractive model for life and work. The description of Norwegian society by Charles Emmerson may stand somewhat *pars pro toto* for life in the EHN:

> Some Norwegians lament that the principles of social and political consensus are too restrictive. Others argue that stability has been bought at the price of excessive taxation. But from the outside there's something approaching Platonic perfection about it all. There is an elegant, almost classical, equilibrium between freedom and the state, between capitalism and social cohesion, and between national identity and a typically Nordic internationalist vocation. Local affinities – something Plato would have seen as the basis of any true democracy – are still strong, partly because the country's geography conspires against centralisation and partly because the country's history saw a constant shift of power between its cities. [...] Society is made up of closely interlocking professional, social and regional networks. Shared values predominate. Even now, the population of the country is barely 5 million.
>
> (Emmerson, 2011: 272 et seq.)

He followed up with the obvious: "Above all, Norway is rich" (Emmerson, 2011: 273). The latter does not apply in equal measure to Sweden and Finland, but Emmerson's cultural observations might also have been made of one of the other Nordic countries.

Integrating into Nordic societies, in particular in the sparsely populated EHN, requires a certain mindset, which might be more difficult to acquire than local language skills. Looking at the way schools in the EHN instill values such as *sisu* – a trait that is considered typically Finnish and might be explained (rather than translated) as "grit", "strong will in the face of adversity" or "endurance" – in children, it becomes clear that this mindset is not country-specific and that it can be acquired by immigrants, too. Societies in the EHN are about freedom, equality and mutual respect, which is expressed in allowing each other space while remaining an active part of society. The focus on the strong welfare state, which is a predominant feature of the Nordic countries when seen from the outside, should not distract from the role of civil society in the EHN. Although not necessarily as formalized as, for example, in Germany, where the *Vereinsleben* – active membership of associations – is an important aspect of societal life in towns and villages, civil society plays an important role in the EHN, too. Therefore, also in the EHN, integration of migrants first and foremost happens at the local level, in the workplace and neighbourhoods. Adopting this Nordic understanding of freedom and a willingness to share – the latter being reflected in the wide acceptance of very high tax rates and the resulting level of post-taxation income equality compared with other states – is essential in order to thrive in the EHN. Often, this mindset also facilitates economic and professional success in the Nordic countries, although discrimination and inequalities are very real challenges faced (not only) by immigrants and other minorities. Although imagined ideas of a strong welfare state might dominate the mental picture outsiders might have of the North, practical values, such as *sisu* (on the "Nordic", as opposed to merely Finnish, character of *sisu*, see Pantzar, 2018: 40) and the corresponding work ethic, are success factors for integration that should not be underestimated. For many migrants who have moved to the EHN and who already possessed these values prior to their arrival, the EHN has indeed proven to be a place of many opportunities. From the outside, the Nordic countries are perceived as almost a paradise (see e.g. Booth, 2014, or Stokowski, 2019), an assessment that is often based on personal, local experiences. Local effects can be put into a wider context: not only are municipalities in the Nordic countries relatively independent from the state (Martínez Soria, 2007: 1018 et seq.), it can be said that,

> [t]he state exists as a tax collector, but the money is spent in the communes themselves, directed by the communes – for, say, skills training locally determined as deemed necessary by the community themselves, to respond to private demand for workers. The economic elites have more freedom [in the Nordic countries] than in most other democracies – this is far from the statism one can assume from the outside.
>
> (Taleb, 2013: 131)

And this helps local communities build up resilience against economic shocks (ibid.).

In September 2019, it was reported that the municipalities of Kemi, Rovaniemi, Salla, Tervola, Tornio and Ylitornio were set to take in more refugees (Yle, 2019b). This decision came on the back of a rather successful integration of refugees, especially in smaller, remote municipalities that one might not immediately suspect had the capacity to do so. Approximately half of the refugees taken in by Salla in the years 2017–2019 have remained in the small, remote municipality (Yle, 2019b), which is located on the border with Russia. Another success story can be seen in the integration of refugees on the western border with Finnish Lapland, where, in the municipality of Ylitornio, which borders Sweden, no refugees have moved away from the municipality (Yle, 2019a), thanks to the integration efforts made locally. Among the factors facilitating the integration of migrants, including refugees, in Finnish Lapland has been the involvement of existing migrant communities in the region (Yle, 2018). From the perspective of municipalities, the integration of refugees has already been economically beneficial owing to the financial support provided by the state (Kourilehto, 2019), and it has even been suggested by Schwan and Zobel that funding for municipalities for the integration of refugees should be provided by the European Union itself (Schwan & Zobel, 2019).

Migrants and local economic development

Immigrants are said to find it difficult to obtain work in the EHN (for Finnish Lapland, see Yeasmin, 2018: 150), although employer attitudes are said to differ with regard to immigrants from, for example, Europe, when compared with immigrants from developing countries (Yeasmin, 2018: 151). It has to be noted, therefore, that migrants often face discrimination in the Nordic countries. In that, they are not alone. Despite thousands of years of interaction (Byers, 2014: 216) and hundreds of years of legal relationships (Byers, 2014: 217), dating back at least to the 1751 Lapp Codicil (Byers, 2014: 217.), discrimination against indigenous Sámi persons continues to this day. The societies of the EHN are not as ethnically homogeneous as they are often perceived, but the societal structure can make it difficult for migrants to integrate into the local societies. Being able to work and contributing to the society, both in general and at the local level, are particularly important factors when it comes to the integration of migrants into Nordic societies. There is, therefore, a link between migration and economic development. This link exceeds the idea of (temporary) work migration for the purpose of filling gaps in the labour market – for example, seasonal work in the tourism, hospitality or agricultural industries.

Such a rejuvenation of the economy might often be helpful at the local level: often, economic development in the EHN is seen primarily through the lens of the extractive industries (Expert Center for Arctic Development PORA, 2019: 26 et seq.) or of the tourism industry. The ageing societies of the EHN can benefit from migration not only in human, but also in social, terms, and, by becoming a home for new people, Arctic locations import not only workers and consumers, but also new ideas. Migration can help smaller municipalities to maintain population numbers and to prevent a further loss of people and infrastructure, such as schools (Schwan & Zobel, 2019). Smaller rural communities that actively integrate migrants might even have the potential to become more attractive destinations for migrants than larger cities (Schwan & Zobel, 2019), although cities have long been the preferred destination for migrants owing to better job opportunities and, often, the presence of established immigrant communities. In a sense, moving to a smaller municipality increases the pressure on migrants to actively integrate into the host society at a local level. The integration of refugees can be economically beneficial for municipalities, if the integration is undertaken on the local – that is, municipal – level, if it involves actual volunteers, and if it is connected to more investment in the municipalities in question (Schwan & Zobel, 2019a).

It would seem unrealistic to burden refugees, who have been forced to abandon their homelands and are now trying to integrate into a new society under very different cultural, social and climatic conditions, with overcoming demographic trends that have threatened the long-term welfare of societies in the EHN for some time. However, migrants can and do make significant economic and social contributions to the host societies. In order to be able to do so, they need the support of the local communities and the tools that are necessary to thrive in these societies. Often, this will mean access to meaningful education.

Human rights aspects of sustainable development

Emphasizing the sustainability of the economic development of the far North is important not only in terms of long-term economic planning and the protection of the natural environment, but also in legal terms. Lawyers navigate a different kind of environment, one of regulations. From the outside, legal limits on the work of corporations are often perceived as limitations, whereas, from the lawyer's perspective, they often are simply different doors to opportunities.

The sustainability of economic development, the availability of multiple income options, is especially important for smaller towns and villages in the Arctic. Without realistic job prospects, populations will dwindle, and a way of life will come to an end, as has already happened in many communities. Given the abundance of jobs that can be done from virtually anywhere, this is no longer an unchangeable fate. This is also realized when we look at the Sustainable Development Goals of the United Nations (2015). The Sustainable Development Goals are human-centred. Most goals are not abstract but aim to speak directly to human needs – for example, Sustainable Development Goals 3 ("Good Health"), 4 ("Quality Education") and 6 ("Clean Water"). It is here where we find the link between sustainable development and human rights. Education is a human right that has been recognized in a number of international treaties – for example, in Article 13 of the International Covenant on Economic, Social and Cultural Rights (ICESCR), which applies almost everywhere in the Arctic (except in Alaska). Likewise, Article 12 of the same treaty guarantees a right to health. Human health is inextricably linked to the environment in which we live. In the last two decades, even the European Court of Human Rights has widely recognized the right to a healthy environment under Article 8 of the European Convention on Human Rights (Council of Europe, 1950), recognizing the duty of the state to protect the individual against unhealthy effects of environmental pollution. International human rights treaties, therefore, can lead to higher levels of environmental protection at the national and local level. Therefore, we ought no longer to look at legal restrictions on corporate activities as bureaucratic red tape, but should keep in mind the purpose and origin in particular of environmental law standards.

But this is only one part of the equation. Businesses ought to do well, in a responsible, sustainable manner, because, in doing so, in providing jobs and direct as well as indirect income, they contribute to the realization of an additional human right, the human right to development. As early as 1986, the General Assembly of the United Nations adopted the Declaration on the Right to Development (United Nations, 1986). Today, this declaration contains important ideas for the understanding of sustainable development in the context of human rights:

> 1. The right to development is an inalienable human right by virtue of which every human person and all peoples are entitled to participate in, contribute to, and enjoy economic, social, cultural and political development, in which all human rights and fundamental freedoms can be fully realized.
>
> (United Nations, 1986, Article 1 (1))

This right "implies the full realization of the right of peoples to self-determination" (Article 1 (2)), but, most importantly, "[t]he human person is the central subject of development and should be the active participant and beneficiary of the right to development" (Article 2 (1)). This role as an "active participant" combines rights and duties: we all

> have a responsibility for development, individually and collectively, taking into account the need for full respect for [our] human rights and fundamental freedoms as well as their duties to the community, which alone can ensure the free and complete fulfilment of the human being.
>
> (United Nations, 1986, Article 2 (2))

Here, too, special attention is given to the issues of education and human health (Article 8 (1) 1)).

The human rights dimension of sustainable development is characterized by its social nature. Economic development is not an end in itself, but serves the community and the human person. The human being is at the core of both international human rights law and sustainable development. Especially under conditions of rapid change, it is important to remember that the individual is supposed to play an active role in the sustainable development of the economy. In the Arctic, this means a specific emphasis on education beyond short-term workforce requirements and creating meaningful work options, including in rural areas. This is a field that offers itself up for cross-border cooperation and the use of modern forms of communication.

Education is a human right (Smith, 2014: 334); indeed, Article 26 paragraph 1 of the Universal Declaration of Human Rights (UDHR; United Nations, 1948) requires elementary education to be "compulsory" (Article 26 paragraph 1 sentence 3) and free of charge (Article 26 paragraph 1 sentence 2 UDHR; see Smith, 2014: 334). Higher education is supposed to "be equally accessible to all on the basis of merit" (Article 26 paragraph 1 sentence 4 UDHR). Although the international human rights law does not require secondary education to be free of charge (Smith, 2014: 335), the International Covenant on Economic Social and Cultural Rights (United Nations, 1966) and the Convention on the Rights of the Child (CRC; United Nations, 1989) oblige states to move towards the removal of financial considerations as a barrier to free higher education (see Smith, 2014: 335). Other barriers, such as discrimination (see Smith, 2014: 335 et seq.), should also be removed, so as to make education as widely accessible as possible. Already under existing international law, basic education has to be available to all, and "[t]he sole ground for access to higher education should be merit" (Smith, 2014: 337). This approach is based on the assumption of equal access to basic education and a fair chance for everybody to achieve the merit required for more education, which cannot be taken for granted everywhere. The practical realization of the right to education, in particular access to higher education, can go a long way towards the creation of conditions that also allow economic development in extremely remote areas to become sustainable. It is, therefore, hardly surprising that education is one of the United Nations' Sustainable Development Goals.

Migrants are in particular need of access to meaningful education that enables them to participate actively in the cultural life of the society (itself a right under Article 15 ICESCR) and to access the labour market. Job-related education, such as vocational training, takes a special place in international human rights law in that it is not only covered by the right to education under Article 13 ICESCR, but is also closely linked to the right to work (Article 6 (1) ICESCR) by inclusion in Article 6 (2) ICESCR. Finland is particularly noteworthy for its efforts to make higher education accessible also through non-traditional approaches (Yle, 2019c). Access to education matters for the communities in the EHN because education is the key to regional development that is sustainable in the long run, instead of relying on short-term economic gains. Right now, the Arctic is often seen as a place from where resources can be taken or that can be used – often at the expense of the natural environment of the Arctic and of the people who live there. High-quality education and modern technology, on the other hand, can contribute to ensuring sustainable development of the Arctic, allowing people to live, study and work in or near their communities. That young people leave rural areas to search for educational opportunities or work in the cities is not a new phenomenon, but the effects are felt particularly in small communities in the Arctic. The idea behind Goal 4 of the UN Sustainable Development Goals is to "[e]nsure inclusive and equitable quality education and promote lifelong learning opportunities for all".

High-quality education is essential for moving the Arctic towards more sustainable forms of development. This includes not only looking for innovative alternative forms of income, such as data centres, which bring new opportunities but also new problems in terms of land and energy requirements, but also searching for ways to make existing business activities, such as tourism, less harmful. At the same time, education is necessary in order to raise awareness of specific Arctic needs among investors and outside actors, and also awareness of rights among the people in the Arctic who might be affected by new business ventures.

Concluding remarks

For Europe, the far North is a remote periphery (cf. Marshall, 2016: 92), and it is hardly surprising that the notion of migration to this cold part of the continent is rarely a concern outside the region. For small communities that suffer from declining population numbers, migration can provide a lifeline and hope for the future continuation of the local society, hope that the local community's story will not come to an end just yet.

The greatest treasure of the lands in the far North is not the natural resources, not the oil or gas, it is not even the beauty of our nature, but it is the people who live there. Truly sustainable economic development of the far North, therefore, has to focus on the people, in particular on the future generations. Integration into a new society has to happen at the local level. This focus on municipalities also appears justified when one looks at small-scale local communities from a long-term historic perspective, as villages and municipalities, at least in Europe, have often endured much longer than nation-states. It is, therefore, hardly surprising that Nassim N. Taleb sees potential for "City-states and small corporations" (Taleb, 2013: 331) as opposed to the declining importance of, for example, nation-states, thereby providing another indicator for the importance of the local level for governance. Municipalities, however, have made it clear that successful integration depends on jobs, which mostly can be provided not by the public sector, but by commercial employers; municipalities, therefore, correctly expect corporations to do their part, too (dpa/epd, 2017). Integration through work in turn requires that the skill sets offered by migrants actually match the needs of employers. Although this is less of a problem in the case of planned migration for the purpose of taking up a specific job in another country, it remains a problem for refugees who had not planned to leave their home country and often had hardly any choice with regard to the exact place where they actually ended up living. A mismatch between present and required skill sets then highlights the need for the accessibility of training and education for migrants. Education is of crucial importance for the success of integration measures.

With regard to the competences and the financing received by municipalities, but also owing to the strong focus on the importance of education, local communities in the EHN should be well equipped to contribute to the effective integration of migrants. What is needed is the active participation of everybody concerned, both long-term residents and migrants, as well as the political will in the power centres to provide adequate support to those who are engaged in the daily work of integrating migrants in the periphery, in some of the most sparsely populated and challenging environments in Europe.

References

Aumüller, Jutta (2018). "Die kommunale Integration von Flüchtlingen", in: Gesemann, Frank & Roth, Roland (eds.), *Handbuch lokale Integrationspolitik*. Wiesbaden: Springer, pp. 173–198.

Aumüller, Jutta & Bretl, Carolin (2008). *Lokale Gesellschaften und Flüchtlinge: förderung von sozialer Integration - Die kommunale Integration von Flüchtlingen in Deutschland*. Berlin: Berliner Institut für Vergleichende Sozialforschung.

Bederman, David J. (2008). *Globalization and International Law*. New York: Palgrave Macmillan.

Bertelsmann Stiftung (2018). *Ankommen in Deutschland - Geflüchtete in der Kommune integrieren – Ein Handbuch*. Gütersloh: Bertelsmann Stiftung. www.bertelsmann-stiftung.de/en/publications/publication/did/online-handbuch-ankommen-in-deutschland-1/ (accessed 1 October 2019).

Booth, Michael (2014). *The Almost Nearly Perfect People – Behind the Myth of the Scandinavian Utopia*. London: Jonathan Cape.

Byers, Michael (2014). *International Law and the Arctic*. Cambridge: Cambridge University Press.

Convention on the Rights of the Child (1989). "United Nations Convention on the Rights of the Child, United Nations Treaty Series 1577, 3", https://treaties.un.org/doc/Treaties/1990/09/19900902%2003-14%20AM/Ch_IV_11p.pdf (accessed 2 October 2019).

Declaration on the Right to Development (1986). "United Nations General Assembly, United Nations General Assembly Resolution A/RES/41/128", www.ohchr.org/EN/ProfessionalInterest/Pages/RightToDevelopment.aspx (accessed 2 October 2019).

Der Beauftragte der Bundesregierung für Migration, Flüchtlinge und Integration. (2019). "Kommunen können Integration", www.integrationsbeauftragte.de/ib-de/themen/gesellschaft-und-teilhabe/vor-ort/-kommunen-koennen-integration–410786 (accessed 1 October 2019).

Deutscher Städtetag. (2016). *Flüchtlinge vor Ort in die Gesellschaft integrieren - Anforderungen für Kommunen und Lösungsansätze*. Berlin/Cologne: Deutscher Städtetag.

Deutsches Institut für Urbanistik. (2016). "Flüchtlinge und Asylsuchende in Kommunen", https://difu.de/10258 (accessed 30 September 2019).

dpa/epd (2017). "Kommunen fordern mehr Anstrengungen auch von Unternehmen", in: *Frankfurter Allgemeine Zeitung*, 25 December 2017, https://faz.net/aktuell/politik/fluechtlinge-kommunen-wollen-bessere-integration-15357749.html (accessed 1 October 2019).

Durfee, Mary & Johnstone, Rachael Lorna (2019). *Arctic Governance in a Changing World*. Lanham/Boulder/New York/London: Rowman & Littlefield.

Emmerson, Charles (2011). *The Future History of the Arctic*. London: Vintage Books.

European Convention on Human Rights (1950). "Convention for the Protection of Human Rights and Fundamental Freedoms", www.echr.coe.int/Documents/Convention_ENG.pdf (accessed 2 October 2019).

Expert Center for Arctic Development PORA. (2019). *Polar Index of the Barents Region – Sustainable Development Ratings of Provinces and Companies*. Moscow: Lomonosov Moscow State University, Faculty of Economics, Environmental Economic Department.

Heinämäki, Leena (2010). *The Right to Be a Part of Nature – Indigenous Peoples and the Environment* (1st ed.). Rovaniemi: Lapland University Press.

Heisterhagen, Nils (2019). "Es braucht kein Dorf - Die Verteilung von Flüchtlingen über die Kommunen führt in die Irre. Die Entscheidung muss beim Nationalstaat bleiben", in: *Internationale Politik und Gesellschaft*, 11 September 2019, www.ipg-journal.de/schwerpunkt-des-monats/europaeische-asylpolitik/artikel/detail/es-braucht-kein-dorf-3714/ (accessed 1 October 2019).

Hessische Staatskanzlei. (2018a). "Förderprogramme für ehrenamtliche Flüchtlingshilfe wird fortgesetzt", 23 April 2018, https://staatskanzlei.hessen.de/pressearchiv/pressemitteilung/foerderprogramm-fuer-ehrenamtliche-fluechtlingshilfe-fortgesetzt-0 (accessed 30 September 2019).

Hessische Staatskanzlei. (2018b). *Integration von Flüchtlingen im ländlichen Raum – Abschlussdokumentation*. Wiesbaden: Hessische Staatskanzlei.

Hilton, Steve (2016). *More Human – Designing a World Where People Come First*. London: WH Allen.

International Covenant on Economic Social and Cultural Rights (1966). "United Nations Treaty Series 993, 3", https://treaties.un.org/doc/Treaties/1976/01/19760103%2009-57%20PM/Ch_IV_03.pdf (accessed 2 October 2019).

Karakurt, Türkan (no year). "Warum kann Integration nur in den Kommunen gelingen?", in: *Grundwissen der Kommunalpolitik Baden-Württemberg 17*. Bonn: Friedrich-Ebert-Stiftung, pp. 2–5.

Kirchner, Stefan (2018). "Indigenous Rights and Livelihoods as Concerns in the Decision-Making on Extractive Industries in Finland", in Hossain, Kamrul, Roncero Martín, José Miguel, & Petrétei, Anna (Eds.), *Human and Societal Security in the Circumpolar Arctic – Local and Indigenous Communities*. Leiden: Brill Nijhoff, pp. 263–280.

Kommunale Gemeinschaftsstelle für Verwaltungsmanagement. (2017a). "Kommunales Integrationsmanagement. Teil 1: Managementansätze und strategische Konzeptionierung (7/2017)", www.bertelsmann-stiftung.de/fileadmin/files/Projekte/Ankommen_in_Deutschland/B7_2017_Kommunales-Integrationsmanagment_Teil_1.pdf (accessed 2 April 2020).

Kommunale Gemeinschaftsstelle für Verwaltungsmanagement. (2017b). "Kommunales Integrationsmanagement. Teil 2: handlungsfelder und Erfolgsfaktoren gestalten (15/2017)", www.kgst.de/dokumentdetails?path=/documents/20181/1379003/15-2017-Kommunales-Intergrationsmanagement-Teil-2/64fa4603-957f-738e-d013-a3e55767d7a3 (accessed 1 October 2019).

Kourilehto, Tuulikki (2019). "Ylitornion kunnan alijäämä kutistui - pakolaisista maksetut korvaukset ja hankeet toivat tuloja ennakoitua ennemän", in: *Lapin kansa*, 24 June 2019, https://lapinkansa.fi/ylitornion-kunna-alijaama-kutistui-pakolaisista-m/170838 (accessed 1 October 2019).

Krakauer, Jon (2007). *Into the Wild*. London: Pan Books.

Landeszentrale für politische Bildung Baden-Württemberg. (2019). "Flüchtlinge und Schutzsuchende in Deutschland", https://lpb-bw.de/fluechtlingsproblematik.html (accessed 1 October 2019).

Lehmann, Julian (2019). "Flucht nach vorn? Bei der Reform ihres Asylsystems hat sich die EU in eine Sackgasse manövriert. Von der Leyen will einen Neubeginn", in: *Internationale Politik und Gesellschaft*, 27 August 2019, www.ipg-journal.de/schwerpunkt-des-monats/europaeische-asylpolitik/artikel/detail/flucht-nach-vorn-3689/ (accessed 1 October 2019).

Lopez, Barry (2014). *Arctic Dreams [Reissue], With an Introduction by Robert Macfarlane*. London: Vintage Books.

Marshall, Tim (2016). *Prisoners of Geography (paperback ed.)*. London: Elliot & Thompson.

Martínez Soria, José (2007). "§ 36 Kommunale Selbstverwaltung im europäischen Vergleich", in: Mann, Thomas & Püttner, Günter (Eds.), *Handbuch der kommunalen Wissenschaft und Praxis*. Berlin/Heidelberg/New York: Springer, pp. 1015–1043.

Ohliger, Rainer, Schweiger, Raphaela, & Veyhl, Lisa (no year). *Auf dem Weg zur Flüchtlingsintegration in ländlichen Räumen: ergebnisse einer Bedarfsanalyse in sieben Landkreisen*. Stuttgart: Robert Bosch Stiftung.

Outwater, Alice (1996). *Water – A Natural History* (1st ed.). New York: Basic Books.

Pantzar, Katja (2018). *Finding Sisu – In Search of Courage, Strength and Happiness the Finnish Way*. London: Hodder & Stoughton.

Schwan, Gesine & Zobel, Malisa (2019). "Asylblockade lösen, Gemeinden stärken - Städte und Kommunen sollten selbst über die Aufnahme von Geflüchteten entscheiden. Davon könnten sie mehrfach profitieren", in: *Internationale Politik und Gesellschaft*, 9 September 2019, www.ipg-journal.de/rubriken/europaeische-integration/artikel/asylblockade-loesen-gemeinden-staerken-3703/ (accessed 1 October 2019).

Schwan, Gesine & Zobel, Malisa (2019a). "Es braucht ein ganzes Dorf! Die freiwillige Aufnahme von Flüchtlingen in europäischen Kommunen sorgt für eine demokratische Wiederbelebung von unten", in: *Internationale Politik und Gesellschaft*, 24 September 2019, www.ipg-journal.de/schwerpunkt-des-monats/europaeische-asylpolitik/artikel/detail/es-braucht-ein-ganzes-dorf-3746/ (accessed 1 October 2019).

Smith, Rhona K. M. (2014). *Textbook on International Human Rights* (6th ed.). Oxford: Oxford University Press.

Stokowski, Margarete (2019). "Besser als woanders", in: *SPIEGEL Online*, 1 October 2019, www.spiegel.de/kultur/gesellschaft/finnland-das-paradies-im-norden-europas-kolumne-a-1288406.html (accessed 1 October 2019).

Sweden. (2011). "Sweden's Strategy for the Arctic Region", www.government.se/49b746/contentassets/85de9103bbbe4373b55eddd7f71608da/swedens-strategy-for-the-arctic-region (accessed 17 September 2019).

Taleb, Nassim Nicholas (2013). *Antifragile – Things that Gain from Disorder*. London: Penguin Books.

United Nations Sustainable Development Goals (2015), "United Nations General Assembly, Transforming Our World: the 2030 Agenda for Sustainable Develpoment, United Nations General Assembly Resolution A/RES/70/1", https://sustainabledevelopment.un.org/content/documents/21252030%20Agenda%20for%20Sustainable%20Development%20web.pdf (accessed 2 October 2019).

Universal Declaration of Human Rights (1948). "United Nations General Assembly Resolution 217 (III)," 10 December 1948, https://undocs.org/A/RES/217(III) (accessed 17 September 2019).

Westra, Laura (2008). *Environmental Justice and the Rights of Indigenous Peoples, International and Domestic Legal Perspectives* (1st ed.). Abingdon: Earthscan.

Yeasmin, Nafisa (2018). *The Governance of Immigration Manifests Itself in Those Who Are Being Governed*. Rovaniemi: Lapland University Press.

Yle (2018). "Lappin on tulossa sata kiintipakolaista syyskuussapakolaisten kotoutumista edistetään keräyksilla ja vahvistamalla syyrialaisten yhteissöä", in: *Yle Uutiset*, https://yle.fi/uutiset/3-10347250 (accessed 1 October 2019).

Yle (2019a). "Ylitornion kunta on onnistunut kiintiöpakolaisten kotouttamisessa poismuuttajien saldo on nolla", in: *Yle Uutiset*, 4 September 2017, https://yle.fi/uutiset/3-10952763 (accessed 1 October 2019).

Yle (2019b). "Salla haluaa ottaa lisää pakolaisiä", in: *Yle Uutiset*, 17 September 2019, https://yle.fi/uutiset/3-10974464 (accessed 1 October 2019).

Yle (2019c). "New pathway to admission: finnish universities expand free online course selection", in: *Yle News*, 28 September 2019, https://yle.fi/uutiset/osasto/news/new_pathway_to_admission_finnish_universities_expand_free_online_course_selection/10995162 (accessed 30 September 2019).

Index

Printed in the United States
by Baker & Taylor Publisher Services